Structural Engineering

Structural Engineering

Edited by Ray Anton

CLANRYE
INTERNATIONAL
www.clanryeinternational.com

Clanrye International,
750 Third Avenue, 9th Floor,
New York, NY 10017, USA

ISBN: 978-1-63240-585-2

Cataloging-in-publication Data

Structural engineering / edited by Ray Anton
    p. cm.
Includes bibliographical references and index.
ISBN 978-1-63240-585-2
1. Structural engineering. 2. Engineering. 3. Structural analysis (Engineering). I. Anton, Ray.
TA633 .S77 2017
624.1--dc23

For information on all Clanrye International publications
visit our website at www.clanryeinternational.com

Printed in the United States of America.

Contents

Preface

The ever growing need of advanced technology is the reason that has fueled the research in the field of advanced structural engineering in recent times. It is an important field of engineering concerned with the stability, strength and rigidity of buildings. This book covers in detail some existent theories and innovative concepts revolving around this discipline. The text aims to highlight some of the major specializations in this area by encompassing related fields like earthquake engineering, façade engineering, etc. The topics covered in this book offer the readers new insights in the field of advanced structural engineering. It includes contributions of experts and scientists which will provide innovative insights into this field. Students, researchers, experts and anyone else associated with the discipline of structural engineering will benefit alike from this book.

This book unites the global concepts and researches in an organized manner for a comprehensive understanding of the subject. It is a ripe text for all researchers, students, scientists or anyone else who is interested in acquiring a better knowledge of this dynamic field.

I extend my sincere thanks to the contributors for such eloquent research chapters. Finally, I thank my family for being a source of support and help.

Editor

Seismic fragility analysis of typical pre-1990 bridges due to near- and far-field ground motions

Araliya Mosleh[1] · Mehran S. Razzaghi[2] · José Jara[3] · Humberto Varum[4]

Abstract Bridge damages during the past earthquakes caused several physical and economic impacts to transportation systems. Many of the existing bridges in earthquake prone areas are pre-1990 bridges and were designed with out of date regulation codes. The occurrences of strong motions in different parts of the world show every year the vulnerability of these structures. Nonlinear dynamic time history analyses were conducted to assess the seismic vulnerability of typical pre-1990 bridges. A family of existing concrete bridge representative of the most common bridges in the highway system in Iran is studied. The seismic demand consists in a set of far-field and near-field strong motions to evaluate the likelihood of exceeding the seismic capacity of the mentioned bridges. The peak ground accelerations (PGAs) were scaled and applied incrementally to the 3D models to evaluate the seismic performance of the bridges. The superstructure was assumed to remain elastic and the nonlinear behavior in piers was modeled by assigning plastic hinges in columns. In this study the displacement ductility and the PGA are selected as a seismic performance indicator and intensity measure, respectively. The results show that pre-1990 bridges subjected to near-fault ground motions reach minor and moderate damage states.

Keywords Concrete bridges · Seismic vulnerability · Time history analysis · Fragility curves · Far-field · Near-fault

Introduction

Bridges are important components of transportation systems. Bridge failures due to extreme loading conditions such as earthquakes may cause serious impacts to transportation systems. It is necessary to evaluate the seismic vulnerability of highway bridges to assess the expected economic losses caused by damage to highway systems in the event of an earthquake. There are many different methods to assess bridge performance such as using fragility curves (FC). There are at least four methodologies for the development of seismic fragility curves, namely: expert opinion, empirical, analytical and hybrid approaches (Avsar et al. 2011; Banerjee and Shinozuka 2007; Choine et al. 2015; Mander et al. 2007; Tavares et al. 2012; Yazgan 2015). To obtain the analytical fragility curves three steps should be considered: the simulation of ground motions, the simulation of bridges, and the generation of fragility curves. The nonlinear static analysis (Banerjee and Shinozuka 2007; Dutta and Mander 1998; Loh et al. 2002; Monti and Nistico 2002; Siqueiraa et al. 2014) nonlinear dynamic time history analysis as the most time-consuming and computationally demanding (Shinozuka et al. 2000) and elastic spectral analysis as a simplest and the least time-consuming approach (Hwang et al. 2001) can be evaluate to obtain the structural response. Nielson and

✉ Araliya Mosleh
a_mmosleh@yahoo.com

[1] Department of Civil Engineering, Faculty of Engineering, University of Aveiro, 3810-193 Aveiro, Portugal

[2] Department of Civil Engineering, Faculty of Engineering, Islamic Azad University, Qazvin Branch, Qazvin, Iran

[3] Department of Civil Engineering, Faculty of Engineering, University Michoacana de San Nicolas de Hidalgo, Morelia, Mexico

[4] CONSTRUCT-LESE, Department of Civil Engineering, Faculty of Engineering, University of Porto, 4200-465 Porto, Portugal

DesRoches (2007) proposed the investigation due to the vulnerability of steel and concrete girder bridges by considering nonlinear analyses (Nielson and DesRoches 2007). Choe et al. (2009) studied typical single-bent bridge in California with RC columns, by applying nonlinear static analysis. Bertero et al. (1978) reported the some effects of near-fault ground motion, but they ignored the implications in seismic design. Brown and Saiidi (2008) reported the comparative results of two substandard bridge bents tests under dynamic ground motions on the shaking tables subjected to near-fault and far-field ground motions. The effect of near-fault versus far-field ground motion on beam and column reinforcement was investigated in this study. Several parameters were used to study the effects of near-fault versus far-field ground motions and presented the near-fault caused more extensive apparent damage in the column (Brown and Saiidi 2008). Muntasir Billah et al. (2013) focused on the fragility-based seismic vulnerability assessment of retrofitted multicolumn bridge bents subjected to near-fault and far-field ground motion. Ramanathan et al. (2015) studied the evolution in design details for Californian box-girder bridges. Also, the importance of design details on the fragility of box-girder bridges is quantified in this study. Bridge damages produce both direct and indirect losses that can be extremely high (Padgett and DesRoches 2007). During the past decades several bridges damaged due to the occurrence of earthquakes (Eshghi and Ahari 2005; Eshghi and Razzaghi 2004; Ellingwood et al. 2004; Nicknam et al. 2011; Wang et al. 2009; Wang and Lee 2009; Yang et al. 2015). Hence, the expected seismic performance of bridges attracted several researchers during the last decades (Jara et al. 2011, 2013; Varum et al. 2011; Zhang et al. 2008; Lin et al. 2015).

Before the 1970s, many of the bridges were not designed for withstand earthquakes. During the 1971 San Fernando earthquake in California several bridges suffered damages (Memari et al. 2011). The Loma Prieta earthquake in 1989 caused noticeable damage to bridges. Following the Loma Prieta earthquake, substantial changes have been made to seismic design provisions of the bridges. Seven bridges collapsed during the 1994 Northridge earthquake and many others sustained damages without collapse (Housner and Thiel 1995). Performance of pre-1990 bridges revealed that these structures are seismically vulnerable. The importance of acceptable seismic behavior for bridges in transportation systems has emphasized the need for seismic safety evaluations of existing bridges. In some countries, there is a lack of detailed studies analyzing the seismic vulnerability of the pre-1990 bridges that allows conducting specific tasks to reduce economic losses in the future. Furthermore, fragility curves can incorporate the repair cost and the recovery time for evaluating the seismic performance of a highway system, and the methodology is widely applied to assess the seismic

vulnerability of bridges located in areas of high seismicity (Jara et al. 2012).

The main objective of this study is analyzing the seismic vulnerability of pre-1990 bridges. As a typical bridge structure, one of the most common bridges designed and constructed in the 1980s is selected. The bridge was subjected to a family of seismic records with different dynamic characteristics. Fragility curves were determined for each set of seismic records based on 3D models and nonlinear dynamic time history analyses. The objective of this study is to evaluate the seismic performance of old concrete bridges with different column height in Iran located near and far from sources, by assessing seismic fragility curves. The results allow evaluating the expected seismic performance of the bridges.

Theoretical background

A fundamental requirement for estimating the seismic performance of a particular structure is the ability to quantify the potential for damage as a function of earthquake intensity (e.g., peak ground acceleration). A probabilistic seismic performance analysis (PSPA) based on fragility curves provides a framework to estimate the seismic performance and reliability of the structures (Ellingwood et al. 2004; Razzaghi and Eshghi 2014; Jeon et al. 2015). Fragility functions relate the probability that the demand on a particular structure exceeds its capacity to an earthquake severity measure. It can be expressed as follows:

$$Fr = P[(S_d \geq S_c | SM)] \qquad (1)$$

where Fr = fragility function, S_d = structural demand, S_c = structural capacity and SM = earthquake severity measure. Assuming that the demand and capacity are random variables represented by a standard lognormal function, the Eq. (1) becomes:

$$Fr = P\left[\left(\frac{S_d}{S_c} \geq 1 | SM\right)\right] = \Phi\left[\frac{1}{\beta} \ln\left(\frac{S_d}{S_c}\right)\right] \qquad (2)$$

where $\Phi[.]$ = the standard normal distribution function and β = logarithmic standard deviation of the variables. This assumption has been made by several researchers (Choi et al. 2004; Hancilar et al. 2013; Razzaghi and Eshghi 2014; Shinozuka et al. 2000).

According to Eq. (2), the fragility functions depend on the structural demand and the selected damage states. The structural demand was estimated by conducting nonlinear time history analyses. There are various approaches for establishing damage limit states. HAZUS provides five qualitative damage states varying from no damage to structure collapse, based on the column damages and serviceability of bridges (NIBS 1999) (Table 1). Furthermore

Table 1 Description of bridge damage states, taken from HAZUS (NIBS 1999)

Damage states	Description
No damage (N)	No damage to a bridge
Slight/minor damage (S)	Minor cracking and spalling to the abutment, cracks in shear keys at abutments, minor spalling and cracks at hinges, minor spalling at the column (damage requires no more than cosmetic repair) or minor cracking to the deck
Moderate damage (M)	Any column experiencing moderate cracking and spalling (column structurally still sound), any connection having cracked shear keys or bent bolts, or moderate settlement of the approach
Extensive damage (E)	Any column degrading without collapse (column structurally unsafe), any connection losing some bearing support, or major settlement of the approach
Complete damage (C)	Any column collapsing and connection losing all bearing support, which may lead to imminent deck collapse

several quantitative damage states have been suggested by various researchers based on the strain limit in the column section, crack width and repair cost (Hose et al. 2000; Karim and Yamazaki 2001; Kawashima 2000; Mander 1999). However, the displacement ductility demand is one of the most common quantitative damage parameter used for bridges (Hwang et al. 2001; Mosleh et al. 2015).

In this study, the seismic damage is classified in five damage states, as described by HAZUS (NIBS 1999). To quantify damage states, the relative displacement ductility ratio of a column is used. This variable is defined as:

$$\mu_{ci} = \frac{\Delta_{yi}}{\Delta_{y1}} \tag{3}$$

where μ_{ci} = ductility demand at the ith damage state, Δ_{yi} = relative displacement at the top of a column at the corresponding limit state (i) and Δ_{y1} = relative displacement of a column when the longitudinal reinforcing bars reach the first yield, calculated as follows:

$$\Delta_{y1} = \frac{2}{3}\varphi_{y1}L^2 \tag{4}$$

where L = the length from the plastic hinge to the point of contra-flexure and, φ_{y1} = the curvature corresponded to relative displacement of a column when the vertical reinforcing bars at the bottom of the column reaches the first yield.

Hence, μ_{c1} denotes the first limit state corresponding to a first yield displacement ductility ratio equal to 1. The second damage state, μ_{c2}, represents the yield displacement ductility ratio calculated as:

$$\mu_{c2} = \frac{\Delta_2}{\Delta_{y1}} = \frac{\Delta_y}{\Delta_{y1}} = \frac{2}{3}\frac{\varphi_y L^2}{\Delta_{y1}} \tag{5}$$

where φ_y = the curvature corresponded to relative displacement of a column when the vertical reinforcing bars at the bottom of the column reaches the yield.

The displacement ductility corresponding to the third damage state, μ_{c3}, is the displacement ductility ratio corresponding to $\varepsilon_c = 0.004$; where ε_c is the maximum

compressive strength of concrete column, hence Δ_3 can be estimated by Eq. 6.

$$\Delta_3 = \Delta_2 + \theta_P\left(L - \frac{L_P}{2}\right) \tag{6}$$

where θ_P and L_P are the rotation and the plastic hinge length, respectively. The plastic hinge rotation can be calculated by Eq. 7 and the plastic hinge length can be estimated according to Priestley et al. (1996):

$$\theta_P = (\varphi_3 - \varphi_y)L_P \tag{7}$$

$$L_P = 0.08L + 0.022f_{ye}d_{bl} \geq 0.044f_{ye}d_{bl} \tag{8}$$

where f_{ye} is the yield strength of the reinforcing bars and d_{bl} is the diameter of longitudinal reinforcing bars. Finally μ_{c4} can be calculated as follows (FHWA 1995; Hwang et al. 2001):

$$\mu_{c4} = \mu_{c3} + 3 \tag{9}$$

Ground motion selection

One of the important tasks to generate fragility curves is the correct selection of input motion parameters. The intensity of an earthquake is commonly described using the peak ground acceleration (PGA). However, severe structural damages are not always related with large values of PGA. Other indexes, namely: (PGD) peak ground displacement, (PGV) peak ground velocity, (SI) spectrum intensity (Katayama et al. 1998), (Td) time duration of strong motion (Trifunac and Brady 1975), (D) distance to epicenter, and spectral characteristics, are also employed in damage estimation (Molas and Yamazaki 1995). In this study, analytical probabilistic seismic performance analyses (PSPA) are conducted based on the nonlinear response history of the bridge. All the selected seismic records have PGA greater than 0.05 g (Table 2). This table presents ten strong motions, five of them are near-field and the remaining are far-field records. Near-field ground motions are distinguished by a long period velocity pulse and

Table 2 Important parameters of the selected earthquake ground motions (http://peer.berkeley.edu/peer_ground_motion_database)

Year	D (km)	M	PGA (g)	Earthquake
1989	7.2	6.9	0.644	Loma Prieta
1999	13.7	7.1	0.092	Duzce
1999	33.2	7.4	0.376	Kocaeli, Turkey
1994	40.7	6.7	0.568	Northridge
1978	20.6	7.4	0.406	Tabas
1990	40.4	7.4	0.505	Manjil
1990	84.0	7.4	0.184	Manjil
1987	7.5	6.5	0.793	Superstition Hills
1971	11.8	6.6	0.699	San Fernando
1976	55.7	7.2	0.064	Calderan—Turkey

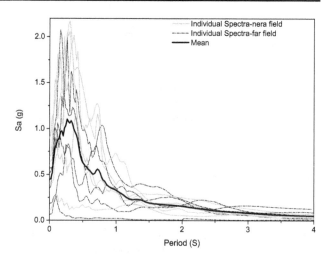

Fig. 1 Response spectra of the selected ground motions

permanent ground displacement (Somerville 2002). Distance to the epicenter of the earthquake is another factor to classify ground motions as 'near-fault', the epicenter should no more than 15 km of the structure. All the seismic stations are located in hard soil sites and the response spectra of the motions are presented in Fig. 1. The natural period of the mentioned bridge sample analyzed is around 2.06 s. It can be notice that the peak in the acceleration response spectra occurs at a period of around 0.5 s. The period of the bridge falls in to the right of the period of the peak response for both near- and far-fields. However, it is observed that the natural period of the response spectra corresponded to far-field earthquakes is more than the mean values, therefor the far-field earthquakes induce a greater acceleration response. Thus, it is expected the mentioned bridge is more vulnerable to seismic effects based on far-field earthquakes.

Description of the bridge

A multi-span simply supported bridge with concrete girders considered a typical structure designed and constructed in the 1980s is selected. The bridge has six spans with a total length of 120 m, and five frame-type bents. Each bent has three circular columns and the superstructure is composed by RC slabs supported on five precast concrete girders spaced at 3.2 m. The span length and the bridge width are 32 and 16 m, respectively. The cap beam is a rectangular element of 1.9 m by 2.0 m and the circular columns have a diameter of 1.4 m. The pier heights are 15 and 18 m. Each column has $20\Phi30$ vertical bars and Φ 18 spiral hoops spacing 200 mm. The gap between deck and abutment is 150 mm and the gap between decks in each span is of 100 mm. The concrete girders are supported on elastomeric type bearings. The geometric characteristics of the bridge are indicated in Fig. 2.

Probabilistic seismic performance analysis

Numerical analysis

The bridges are modeled and analyzed with the SAP2000 software (CSI (SAP2000 V-14) 2009). Frame elements with six degrees of freedom at each node are used to model the columns, bent caps and girders; the deck and diaphragms are modeled with shell elements. Link elements are used to model elastomeric bearings with six degrees of freedom at each node. The nonlinear behavior of the columns is considered with a concentrated plasticity model by assigning plastic hinges at both column ends which is recommended in Caltrans code (Caltrans 2013).

Development of fragility curves

The analytical fragility curves are determined with the results of the response history analyses. The analytical model considers inelastic behavior of the columns and elastic behavior of the deck. The nonlinear time history analysis is carried out by considering the displacement ductility of the columns as limit state. Each fragility curves can be generated as lognormal distribution functions characterized by median and dispersion. Previous studies revealed the noticeable effects of near-fault ground motions on the seismic performance of bridges (Chouw and Hao 2008; Loh et al. 2002; Phan et al. 2007; Taflanidis 2011). To evaluate the importance of the seismic record type on the fragility curves, three curves for each damage state were developed, namely: fragility curves based on near-field ground motions, those developed based on far-field ground motions and fragility curves developed using combination of near- and far-field ground motions. The fragility curves present PGA in the horizontal axis and the

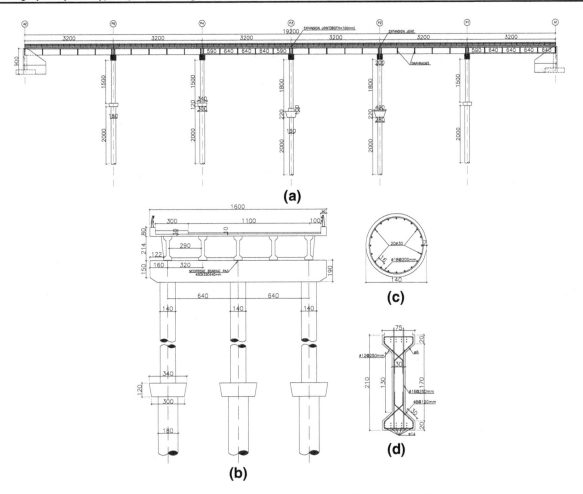

Fig. 2 a Longitudinal view of bridge, **b** bridge cross section, **c** pier cross section, **d** beam cross section

Intensity measure-PGA (g)	LS-1		LS-2		LS-3		LS-4	
	μ	σ	μ	σ	μ	σ	μ	σ
Far	−1.19	0.18	−1.02	0.13	−0.8	0.2	−0.66	0.29
Near	−0.94	0.19	−0.77	0.19	−0.7	0.22	−0.43	0.3
Total	−1.05	0.22	−0.87	0.22	−0.8	0.22	−0.55	0.29

Table 3 Fragility curve parameters for near-field and far-field records

probability of exceedance of a limit state in the vertical axis.

The fragility curves of a bridge display the conditional probability that the structural demand exceeds the structural capacity. Each curve depends on the median value and the dispersion parameter (lognormal standard deviation) of the capacity and the demand. For each ground motion with a specific PGA, the number of sample that reached or exceeded a specified damage limit state is obtained. The probability of exceedance is determined by dividing the number of samples that reached or exceeded the specified damage limit state to the total number of samples. After performing the similar evaluation for each ground motion and the four damage limit states, the

probability of reaching or exceeding the damage limit states is obtained.

Assuming a lognormal density function of the capacity and the demand, the fragility curves are lognormal distributed with the parameters presented in Table 3. These parameters are the results of the nonlinear time history analyses of the bridge subjected to the family of seismic records previously mentioned for each of the limit states.

Figure 3 shows the fragility curves developed based on both near- and far-field ground motions. In this graph LS1, LS2, LS3, and LS4 are: slight, moderate, extensive and complete damage state (collapse), respectively.

If the probability of exceeding a damage performance increased, some mitigation plan should be made in advance. In the limit state of slight damage, minor damage may happen and it is expected that with only small repairs the bridge can be use normally. The difference between the fragility curves for the damage limit states LS2 and LS3 is relatively small. One of the main reasons for this small

difference is the acceptance criteria definitions of the corresponding damage limit states (1.26 corresponded to LS2 and 1.49 related to LS3).

As indicated in Fig. 3, for PGA = 0.5 g (for example) the probability of collapse of the bridge is 30 %. The probability of reaching or exceeding the LS1, LS2 and LS3 are 92, 80, and 68 % respectively. In other words, the fragility curves developed in this study indicate that the selected typical pre-1990 bridges are seismically vulnerable. Figure 4 displays the curves for the two groups of accelerograms and four limit states. Each graph presents three fragility curves to evaluate the effect of the different seismic records on the fragility curves.

The bridge was more vulnerable to the far-field accelerograms, which means that the mentioned bridge is more vulnerable to seismic effects due to far-field earthquakes than near faults. This outcome is consistent with the response of the bridge observed in the earthquakes in far-field. It seems that the bridge location in relation of the epicenter does not impact importantly the fragility curves.

Figure 5 shows the column displacement demands versus the PGA for far-field, near-field, all the seismic records and mead values from elastic behavior through yielding to

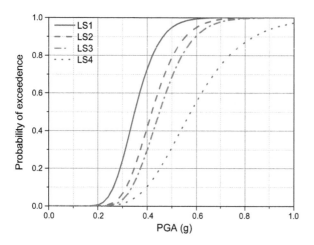

Fig. 3 Bridge system fragility curves for the multi-span simply supported bridge

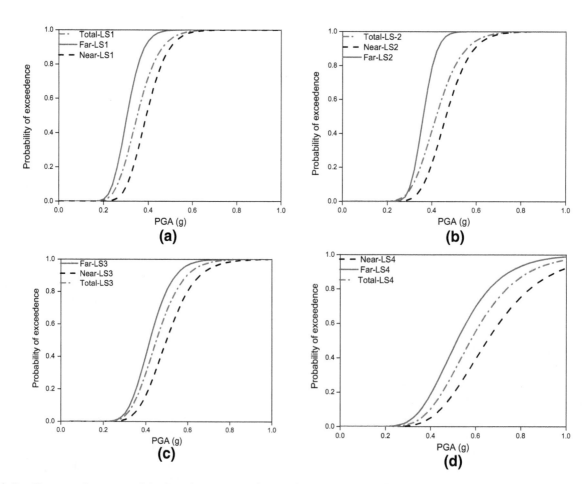

Fig. 4 Fragility curves between **a** slight, **b** moderate, **c** extensive and **d** complete damage limit state

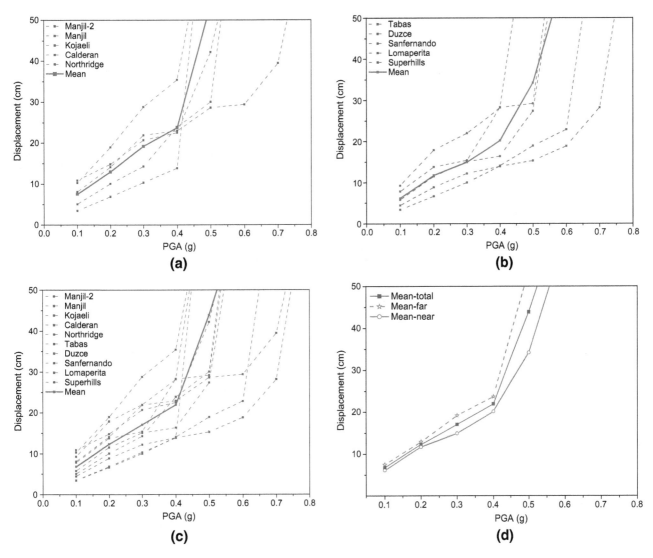

Fig. 5 Incremental displacement **a** far-field, **b** near-field, **c** all seismic records and **d** mean values

dynamic instability or until a limit state failure occurs. Some records produce gradually increments of the displacement demands with the PGA increases and others, like the Kocaeli seismic record, increases suddenly the displacements after a PGA value (0.4 g). For near-field earthquakes, large displacements appears after PGA = 0.5 g in most of the cases. From Fig. 4d it can be observed that the variation of displacement is linear and bridge is in linear zone up to PGA = 0.4 g, while after PGA = 0.5 g the bridge is subjected to nonlinearity.

Conclusions

Fragility curves are a useful tool to estimate the expected damages of bridge structures. The aim of a vulnerability assessment of bridges is to execute preventive actions to plan a disaster response, create a retrofit program, estimate future economic losses, and evaluate the loss of functionality of highway transportation systems. This study presents the generation of fragility curves for one of the most common bridge typologies in high seismic zone areas designed with old codes, and compares the effect of source-to-site distance of group of records. Nonlinear dynamic analyses were carried out to determine the seismic performance of a typical pre-1990 RC bridge subjected to three groups of seismic records: far-field accelerograms, near-field accelerograms and all the seismic records. Earthquake records from some major event e.g., the 1989 Loma perita, the San Fernando 1971, the 1994 Northridge, and some earthquakes recorded in Iran namely: the 1990 Manjil, the 1978 Tabas, were selected as the input ground motions. Analytical fragility curves for the bridge were obtained using the PGA as an intensity measure. The fragility curves for near-fault and far-field sources represent limit states of behavior as a function of the displacement ductility demands on columns. The following conclusions are based on the nonlinear dynamic analyses of the bridge subjected to 70 earthquake records.

- The fragility curves of this study correspond to one of the common class bridges in Iran. It can be used for other type of bridges to determine the seismic risk associated.

- The fragility curves can be used to evaluate potential losses of bridges with the same typology of the analyzed structure. The results showed that far-fields seismic records dominated, while the impact of the near-fields earthquake data bases is reduced.

- Results revealed that the selected typical pre-1990 bridges are seismically vulnerable.

- The columns were the only structural element analyzed in this study; as suggestion for future research the bearings, abutments and the foundation could be included to develop fragility curves.

Acknowledgments The authors would like to thank the anonymous reviewers for their constructive comments.

References

Avsar O, Yakut A, Caner A (2011) Analytical fragility curves for ordinary highway bridges in Turkey. Earthq Spectra 27:971–996

Banerjee S, Shinozuka M (2007) Nonlinear static procedure for seismic vulnerability assessment of bridges. Comput Aided Civil Infrastruct 22:293–305

Bertero VV, Mahin SA, Herrera RA (1978) A seismic design implications of near-fault San Fernando earthquake record. Earthq Eng Struct Dyn 6:21–42

Brown AS, Saiidi MS (2008) Investigation of near-fault vs. far field ground motion effects on a substandard bridge bent. Report University of Nevada, Reno

Caltrans (2013) Seismic design criteria, Version 1.7. California Department of Transportation, Sacramento, CA

Choe D, Gardoni P, Rosowsky D, Haukaas T (2009) Seismic fragility estimates for reinforced concrete bridges subject to corrosion. Struct Saf 31:275–283

Choi E, DesRoches R, Nielson B (2004) Seismic fragility of typical bridges in moderate seismic zones. Eng Struct 26:187–199

Choine M, Connor A, Padgett J (2015) Comparison between the seismic performance of integral and jointed concrete bridges. J Earthq Eng 19:172–191

Chouw N, Hao H (2008) Significance of SSI and nonuniform near-fault ground motions in bridge response: effect on response with conventional expansion joint. Eng Struct 30:141–153

CSI (SAP2000 V-14) (2009) Integrated finite element analysis and design of structures basic analysis reference manual. Computers and Structures Inc., Berkeley, CA

Dutta A, Mander JB (1998) Seismic fragility analysis of highway bridges. Tokyo, Japan, 22–23 June

Ellingwood BR, Rosowsky DV, Li Y, Kim JH (2004) Fragility assessment of light-frame wood construction subjected to wind and earthquake hazards. J Struct Eng. doi:10.1061/(ASCE)0733-9445(2004)130:12(1921)

Eshghi S, Ahari MN (2005) Performance of transportation systems in the 2003 Bam, Iran, earthquake. Earthq Spectra 21:455–468

Eshghi S, Razzaghi MS (2004) The behavior of special structures during the Bam earthquake of 26 December 2003. JSEE Spec Issue Bam Earthq 5:197–207

FHWA (1995) Seismic retrofitting manual for highway bridges publication no. FHWA-RD-94-052. Office of Engineering and Highway Operations R&D, Federal Highway Administration, McLean, VA

Hancilar U, Taucer F, Corbane C (2013) Empirical fragility functions based on remote sensing and field data after the 12 January 2010 Haiti earthquake. Earthq Spectra 29:1275–1310

Hose Y, Silva P, Seible F (2000) Development of a performance evaluation database for concrete bridge components and systems under simulated seismic loads. Earthq Spectra 16:413–442

Housner GW, Thiel CC (1995) The continuing challenge: report on the performance of state bridges in the Northridge earthquake. Earthq Spectra 11:607–636

Hwang H, Liu J, Chiu Y (2001) Seismic fragility analysis of highway bridges. Center for Earthquake Research and Information, The University of Memphis, MAEC RR-4 Project

Jara JM, Galvan A, Jara M, Olmos B (2011) Procedure for determining the seismic vulnerability of an irregular isolated bridge. Struct Infrastruct Eng 9(6):1–13

Jara JM, Jara M, Olmos B, Villanueva D, Varum H (2012) Expected seismic performance of irregular isolated bridges. In: Proceedings of the sixth international IABMAS conference. Bridge maintenance, safety, management, resilience and sustainability, Italy

Jara JM, Jara M, Hernández H, Olmos BA (2013) Use of sliding multirotational devices of an irregular bridge in a zone of high seismicity. KSCE J Civil Eng 17:122–132

Jeon J, Shafieezadeh A, Lee D, Choi E, DesRoches R (2015) Damage assessment of older highway bridges subjected to three-dimensional ground motions: characterization of shear–axial force interaction on seismic fragilities. Eng Struct 87:47–57

Karim KR, Yamazaki F (2001) Effect of earthquake ground motions on fragility curves of highway bridge piers based on numerical simulation. Earthq Eng Struct Dyn 30:1839–1856

Katayama T, Sato N, Saito K (1998) SI-sensor for the identification of destructive earthquake ground motion. Paper presented at the proceedings of the 9th world conference on earth engineering

Kawashima K (2000) Seismic design and retrofit of bridges. Paper presented at the proceedings of 12WCEE, New Zealand

Lin Z, Yan F, Azimi M, Azarmi F, Al-Kaseasbeh Q (2015) A revisit of fatigue performance based welding quality criteria in bridge welding provisions and guidelines. Paper presented at the international industrial informatics and computer engineering conference (IIICEC), Shaanxi, China

Loh CH, Liao WI, Chai JF (2002) Effect of near-fault earthquake on bridges: lessons learned from Chi-Chi earthquake. Earthq Eng Eng Vib 1:86–93

Mander JB (1999) Fragility curve development for assessing the seismic vulnerability of highway bridges. University at Buffalo, State University of New York, MCEER Highway Project/FHWA

Mander J, Dhakal R, Mashiko N, Solberg K (2007) Incremental dynamic analysis applied to seismic financial risk assessment of bridges. Eng Struct 29:2662–2672

Memari AM, Harris HG, Hamid AA (2011) Seismic evaluation of reinforced concrete piers in low to moderate seismic regions. Electron J Struct Eng 11:57–68

Molas GL, Yamazaki F (1995) Neural networks for quick earthquake damage estimation. Earthq Eng Struct Dyn 24:505–516

Monti G, Nistico N (2002) Simple probability-based assessment of bridges under scenario earthquakes. J Bridge Eng 7:104–114

Mosleh A, Varum H, Jara J (2015) A methodology for determining the seismic vulnerability of old concrete highway bridges by using fragility curves. J Struct Eng Geotech 5(1):1–7

Muntasir Billah AHM, Shahria Alam M, Rahman Bhuiyan MA (2013) Fragility analysis of retrofitted multi-column bridge bent subjected to near fault and far field ground motion. J Bridge Eng. doi:10.1061/(ASCE)BE.1943-5592.0000452

NIBS (1999) Vol 2. Earthquake loss methodology, HAZUS 99, Technical manual. Washington DC, USA

Nicknam A, Mosleh A, Hamidi H (2011) Seismic performance evaluation of urban bridge using static nonlinear procedure, case study: Hafez Bridge. Proc Eng 14:2350–2357

Nielson BG, DesRoches R (2007) Analytical seismic fragility curves for typical bridges in the central and southeastern United States. Earthq Spectra 23:615–633

Padgett JE, DesRoches R (2007) Bridge functionality relationships for improved seismic risk assessment of transportation networks. Earthq Spectra 23:115–130

Phan V, Saiidi MS, Anderson J, Ghasemi H (2007) Near-fault ground motion effects on reinforced concrete bridge columns. J Struct Eng 133:982–989

Priestley MJN, Calvi GM (1996) Seismic design and retrofit of bridges. Wiley, New York

Ramanathan K, Padgettb J, DesRoches R (2015) Temporal evolution of seismic fragility curves for concrete box-girder bridges in California. Eng Struct 97:29–46

Razzaghi MS, Eshghi S (2014) Probabilistic seismic safety assessment of precode cylindrical oil tanks. J Perform Construct Facil (ASCE). doi:10.1061/(ASCE)CF.1943-5509.0000669

Shinozuka M, Feng MQ, Lee J, Naganuma T (2000) Statistical analysis of fragility curves. J Eng Mech 126:1224–1231

Siqueiraa G, Sandab A, Paultreb P, Padgettc J (2014) Fragility curves for isolated bridges in eastern Canada using experimental results. Eng Struct 74:311–324

Somerville PG (2002) Characterizing near fault ground motion for the design and evaluation of bridges. Paper presented at the third national conference and workshop on bridges and highways, Portland, Oregon

Taflanidis A (2011) Optimal probabilistic design of seismic dampers for the protection of isolated bridges against near-fault seismic excitations. Eng Struct 33:3496–3508

Tavares DH, Padgett JE, Paultre P (2012) Fragility curves of typical as-built highway bridges in eastern Canada. Eng Struct 40:107–118

Trifunac MD, Brady AG (1975) A study of the duration of strong earthquake ground motion. Bull Seismol Soc Am 65:581–626

Varum H, Sousa R, Delgado W, Fernandes C, Costa A, Jara JM, Álvarez JJ (2011) Comparative structural response of two steel bridges constructed 100 years apart. Struct Infrastruct Eng 7:843–855

Wang Z, Lee GC (2009) A comparative study of bridge damage due to the Wenchuan, Northridge, Loma Prieta and San Fernando earthquakes. Earthq Eng Eng Vib 8:251–261

Wang D, Guo X, Sun Z, Meng Q, Yu D, Li X (2009) Damage to highway bridges during Wenchuan earthquake. J Earthq Eng Eng Vib 3:84–94

Yang C, Werner S, DesRochesa R (2015) Seismic fragility analysis of skewed bridges in the central southeastern United States. Eng Struct 83:116–128

Yazgan U (2015) Empirical seismic fragility assessment with explicit modeling of spatial ground motion variability. Eng Struct 100:479–489

Zhang Y, Conte JP, Yang Z, Elgamal A, Bielak J, Acero G (2008) Two-dimensional nonlinear earthquake response analysis of a bridge-foundation-ground system. Earthq Spectra 24:343–386

2

Optimal placement of active braces by using PSO algorithm in near- and far-field earthquakes

M. Mastali[1] · A. Kheyroddin[2,3] · B. Samali[4] · R. Vahdani[2]

Abstract One of the most important issues in tall buildings is lateral resistance of the load-bearing systems against applied loads such as earthquake, wind and blast. Dual systems comprising core wall systems (single or multi-cell core) and moment-resisting frames are used as resistance systems in tall buildings. In addition to adequate stiffness provided by the dual system, most tall buildings may have to rely on various control systems to reduce the level of unwanted motions stemming from severe dynamic loads. One of the main challenges to effectively control the motion of a structure is limitation in distributing the required control along the structure height optimally. In this paper, concrete shear walls are used as secondary resistance system at three different heights as well as actuators installed in the braces. The optimal actuator positions are found by using optimized PSO algorithm as well as arbitrarily. The control performance of buildings that are equipped and controlled using the PSO algorithm method placement is assessed and compared with arbitrary placement of controllers using both near- and far-field ground motions of Kobe and Chi–Chi earthquakes.

Keywords Pole assignment method · Near- and far-field earthquakes · PSO algorithm · Optimal actuator position

Introduction

The rapid growth of the urban population and consequent pressure on limited space have considerably influenced city residential development. The high cost of land, the desire to avoid ongoing urban sprawl, and the need to preserve important agricultural land have all contributed to drive residential buildings upward. Nowadays, high-rise buildings have become one of the impressive reflections and icons of today's civilization. The outlook of cities all over the world has been changing with these tall and slender structures (Smith and Coull 1991). Tall buildings use load-resisting systems against applied lateral loads such as concrete shear walls and core wall systems. The main problem of these systems is their limitations in controlling the response of super tall buildings (Kheyroddin et al. 2014; Keshavarz et al. 2011). Therefore, some strategies have been used to control and make serviceable tall buildings in addition to the lateral load-resisting systems which included: (1) passive control; (2) semi-active control; (3) active control, or their combinations. In the passive control, the structure uses its internal energy to dissipate external energy. A large number of studies have been conducted on the active control concept (Yang et al. 2004; Kwok et al. 2006). These systems are able to control the structure displacements, accelerations and internal forces by using external energy and providing a direct counter-acting force by the actuators. Using this strategy for controlling structures against external excitation has limitations because of some technological and economic aspects (Symans and Constantinou 1999), as well as the

✉ M. Mastali
m.mastali@civil.uminho.pt; muhammad.mastali@gmail.com

[1] Department of Civil Engineering, ISISE, Minho University, Campus de Azurem, 4800-058 Guimaraes, Portugal

[2] Faculty of Civil Engineering, Semnan University, Seman, Iran

[3] Department of Civil Engineering and Applied Mechanics, University of Texas, Arlington, TX, USA

[4] School of Civil and Environmental Engineering, University of Western Sydney, Sydney, Australia

risks associated with loss of external power in the event of a major earthquake or severe wind load. These systems require high amount of external energy for controlling the structures in comparison with other strategies. Hence, to overcome this problem, semi-active control was proposed as a strategy which compensates for the shortcomings of active control. In this control method, structural properties such as damping and/or stiffness are altered by use of special devices with very little external energy to activate such systems. As a result, much lower amount of external energy is required to control the structures during external excitations and there is a potential in this method to achieve control levels similar to active systems (Amini and Vahdani 2008). Because of limitations in the number of actuators and due to economic reasons, actuator location is an important issue in control problems. Nowadays, numerical methods, such as those inspired by nature, are used in optimizing actuator locations. These methods include ant colony, genetic algorithm, PSO (particle swarm optimization) algorithm, etc.

In this paper, three 3D buildings with different heights are used to investigate the effectiveness of the designed controller. In these systems, the concrete shear walls were also considered as the secondary load-resisting system.

In the present study, three structures with 21, 15, and 9 stories were studied, considering 3, 2, and 1 actuator, respectively. The actuators were placed in the system in two ways, which include (1) finding the optimized position of the optimized actuator using the PSO algorithm; (2) installing an actuator at arbitrary positions. Structures were modelled in MATLAB software. In this regard, finite element method was used for modelling these structures and the interactions between the frame and walls (Ghali et al. 2003). The obtained stiffness from this method was used in modelling of structures in MATLAB software. The novelty of the present paper was using PSO algorithm to optimize actuator locations.

Analysis of a planar frame in the presence of shear walls

The contribution of shear walls in a frame depends on the wall stiffness with respect to other structural elements. Commonly, it is assumed that horizontal forces are applied at the floor levels. Moreover, it is assumed that floor stiffness in the horizontal direction is very high compared to the stiffness of columns and shear walls. Therefore, it is assumed that the floors move as rigid bodies in the horizontal direction. Let us consider the structure shown in Fig. 1, which is constructed of some parallel frames with symmetrical axes.

Some of these structures use shear walls as the secondary load-resisting system. Because of geometric and loading symmetry, floors move without any rotation. By assuming rigid body motion for the floors, each level has a displacement equal to $[D^*]$. The stiffness matrix $[S]_{n \times n}$ (n is the number of stories) corresponding to the coordinate $\{D^*\}$ for each planar frame is calculated and then all the stiffness matrices are assembled to obtain the stiffness matrix of the structure:

$$[S^*] = \sum_{i=1}^{m} [S^*]_i, \tag{1}$$

where m is the number of frames. The lateral movement at the floor level is calculated using the following equation in which $\{F\}$ is the applied force vector on each floor level:

$$[S^*]_{n \times n} \{D^*\}_{n \times 1} = \{F^*\}_{n \times 1}. \tag{2}$$

Approximate analysis of the planar structures

The shear wall deformations are similar to cantilever beams. This simplification is reasonable due to the fact that the rotations are constrained at the ends of the columns by beams in tension (Ghali et al. 2003). It is obvious that shear wall moment of inertia, I, is higher than that of the beam which subsequently leads to reduction in the beam ability to control rotations caused by deformation of the cantilever beam at the floor level. The observed behaviour suggests that the load-resisting systems are composed of two parts; see Figs. 2 and 3.

They are composed of (1) shear wall system; (2) equivalent column. Moment of inertia (I_W) of the shear wall and the column (I_C) at each floor level is equal to the sum of moments of inertia of the shear walls and the columns at that floor level. The second system is an equivalent column which is connected to the beams in a rigid way. Additionally, it is obvious that these two load-resisting systems are connected to each other by non-deformable tension elements and that all the external forces are applied at the floor level. Axial deformations in all structural elements are neglected, while shear deformations in walls and columns can be considered or neglected in the analysis. In case the shear deformations are considered, the effective (reduced) area is equal to the sum of the reduced areas of the walls and the columns at each floor level. It is assumed that the idealized structure has n degrees of freedom, representing the lateral movements of the floors. The stiffness matrix of the structure $\left[S^*_{n \times n}\right]$ is obtained by summing the stiffness matrices of the two resisting systems:

$$[S^*] = [S^*]_w + [S^*]_r, \tag{3}$$

where $[S^*]_r$ and $[S^*]_W$ are the stiffness matrices of the resisting frame and the wall, respectively, corresponding to

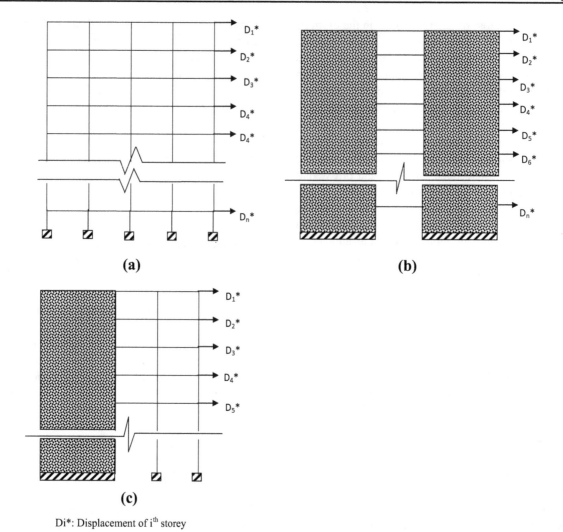

(a)

(b)

(c)

Di*: Displacement of ith storey

Fig. 1 Considered planar frames in the analysis of three dimension and geometrical symmetry: **a** concrete moment resistance frame; **b** symmetric concrete shear walls; **c** asymmetric concrete shear wall (Ghali et al. 2003)

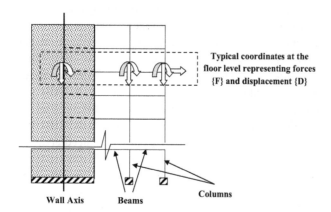

Typical coordinates at the floor level representing forces {F} and displacement {D}

Wall Axis   Beams   Columns

Fig. 2 Coordinate system corresponding to the stiffness matrix $[S]_i$ (Ghali et al. 2003)

n horizontal coordinates at floor levels. To determine $[S^*]_r$ and $[S^*]_W$ matrices, two degrees of freedom are considered at each floor level which consist of a rotation and a lateral movement for the wall and beam–column joint.

According to Fig. 3b, matrices $[S^*]_r$ and $[S^*]_w$ have $2n \times 2n$ degrees of freedom and then these two matrices, which relate the horizontal forces to the lateral movements with non-constrained rotations, are compacted. Therefore, the lateral movement at the floor level is obtained by solving the following equation:

$$[S^*]_{n \times n} \{D^*\}_{n+1} = \{F^*\}_{n \times 1}, \qquad (4)$$

where $\{D^*\}$ represents the horizontal displacements of the shear walls or the columns at each floor level and $\{F^*\}$

Fig. 3 Analysis of simplified structural frame: **a** idealized structure; **b** coordinates corresponding to stiffness matrices $[S_r]$ and $[S_w]$ (Ghali et al. 2003)

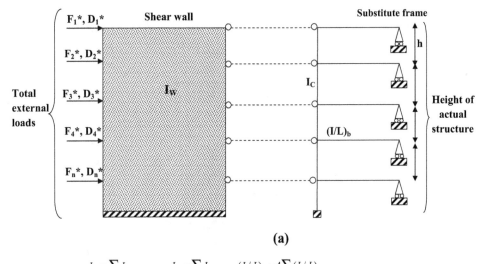

(a)

$$I_w = \sum I_{wi} \qquad I_C = \sum I_{Ci} \qquad (I/L)_b = 4\sum (I/L)_{bi}$$

Subscript of W, C and b refer to wall, column and beam, respectively.

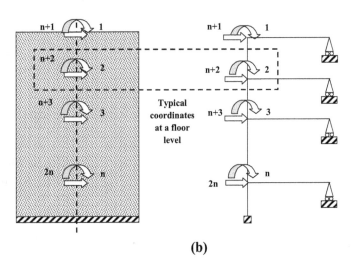

(b)

represents the external loads which are imposed on the shear walls and the columns, i.e. $\{F^*\} = \{F^*\}_w + \{F^*\}_r$.

The applied forces to the shear wall and the frame can be calculated by:

$$\{F^*\}_w = [S^*]_w \{D^*\} \quad \{F^*\}_r = [S^*]_r \{D^*\}. \tag{5}$$

Then these forces are imposed on the shear wall and the frame and, subsequently, moments are determined at the end of the structural elements. If these moments are distributed in the structural elements such as shear walls and frames with regard of their corresponding flexural stiffnesses [(EI/h) or (EI/l)], then the approximate values of the actual end moments can be obtained. It is worth stating that if the shear walls are significantly different, or if there are great variations in the shear wall areas at each floor level, then the above calculation method can lead to false results. In this case, it is necessary to consider an idealized

structure with more than one shear wall connected to the frame by means of tension elements, and the stiffness of each one must be calculated separately. Then the stiffness of the idealized structure can be calculated by summation of the distinct stiffnesses.

Special case: considering the same columns and beams

By considering the same cross-sectional areas for the columns and the beam elements, which leads to equal ratio of $(I/l)_b$ for all the stories, the frame matrix stiffness can be computed by the given equation:

$$[S]_{r_{2n \times 2n}} = \begin{bmatrix} [S_{11}]_r & [S_{12}]_r \\ [S_{21}]_r & [S_{22}]_r \end{bmatrix}, \tag{6}$$

in which the sub-matrices are:

Fig. 4 Plan of structures and shear wall positions

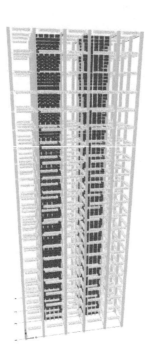

$$[S_{11}]_r = \frac{2(S+t)}{h^2} \cdot \begin{bmatrix} 1 & -1 & 0 & 0 & 0 & 0 & 0 & 0 \\ -1 & 2 & -1 & 0 & 0 & 0 & 0 & 0 \\ 0 & . & 2 & . & 0 & 0 & 0 & 0 \\ 0 & 0 & . & . & . & 0 & 0 & 0 \\ 0 & 0 & 0 & . & . & . & 0 & 0 \\ 0 & 0 & 0 & 0 & 2 & -1 & 0 \\ 0 & 0 & 0 & 0 & 0 & -1 & 2 & -1 \\ 0 & 0 & 0 & 0 & 0 & 0 & -1 & 2 \end{bmatrix}_{n \times n},$$

(7)

$$[S_{21}]_r = [S_{12}]_r^T$$

$$= \frac{(S+t)}{h} \begin{bmatrix} -1 & 1 & 0 & 0 & 0 & 0 & 0 & 0 \\ -1 & 0 & 1 & 0 & 0 & 0 & 0 & 0 \\ 0 & -1 & 0 & 1 & 0 & 0 & 0 & 0 \\ 0 & 0 & . & . & . & 0 & 0 & 0 \\ 0 & 0 & 0 & . & . & . & 0 & 0 \\ 0 & 0 & 0 & 0 & . & . & . & 0 \\ 0 & 0 & 0 & 0 & 0 & -1 & 0 & 1 \\ 0 & 0 & 0 & 0 & 0 & 0 & -1 & 0 \end{bmatrix}_{n \times n},$$

(8)

$$[S_{22}]_r = S \begin{bmatrix} (1+\beta) & c & 0 & 0\;0\;0 & 0 & 0 \\ c & (2+\beta) & c & 0\;0\;0 & 0 & 0 \\ 0 & c & (2+\beta) & c\;0\;0 & 0 & 0 \\ 0 & 0 & c & .\;.\;0 & 0 & 0 \\ 0 & 0 & 0 & .\;.\;. & 0 & 0 \\ 0 & 0 & 0 & 0\;.\;. & . & 0 \\ 0 & 0 & 0 & 0\;0\;c & (2+\beta) & c \\ 0 & 0 & 0 & 0\;0\;0 & c & (2+\beta) \end{bmatrix}_{n \times n},$$

(9)

where:

$$\beta = \frac{3E}{S}\left(I/L\right)_b \quad S = \frac{(4+\alpha)}{(1+\alpha)}\frac{EI_c}{h} \quad t = \frac{(2-\alpha)}{(1+\alpha)}\frac{EI_c}{h}$$

$$c = \pi(t/S),$$

(10)

In which S is the rotational stiffness of a column when the support is clamped; t is the transferred moment and c is the transferred coefficient. The shear deformation of the vertical elements can be calculated by:

$$\alpha = \frac{12EI_c}{h^2 Ga_{rc}}$$

(11)

and the shear deformation of the beams are neglected. The effective cross-sectional area of the columns is computed by adding all the effective cross-sectional areas of the columns ($a_{rc} = \sum a_{rci}$).

Modelling and specifications of the models

In this study, three 3D structures with 9, 15 and 21 stories are modelled. The plan of the structures and the shear wall positions are shown in Fig. 4. Moreover, the considered properties for the models are given in Table 1. Since the concrete shear walls were located in one direction and actuators will be placed in the same direction as the concrete shear walls as in Fig. 4, the simplified planar formulas in "Analysis of a planar frame in the presence of shear walls" could be used for modelling the interaction of frame and shear walls in the modelled three-dimensional structures.

Table 1 Considered properties of structures

	Shear wall	Frame	
a_{rw}	6.66	6.66	Considered equal for all structures
α	12.9	0	
β	0	17.8	
E	$10^7 \times 2236$ (N/m^2)	$10^7 \times 2236$ (N/m^2)	
υ	0.17	0.17	
G	$\left(\frac{1}{2.3}\right)E$	$\left(\frac{1}{2.3}\right)E$	
I_c	28.188 (m^4)	0.022 (m^4)	Considered for structures with 21 stories
$\left(\frac{l}{l}\right)_b$	0.25	0.25	
I_c	28.188 (m^4)	0.02 (m^4)	Considered for structures with 21 stories
$\left(\frac{l}{l}\right)_b$	0.22	0.22	
I_c	28.188 (m^4)	0.017 (m^4)	Considered for structures with 21 stories
$\left(\frac{l}{l}\right)_b$	0.20	0.20	

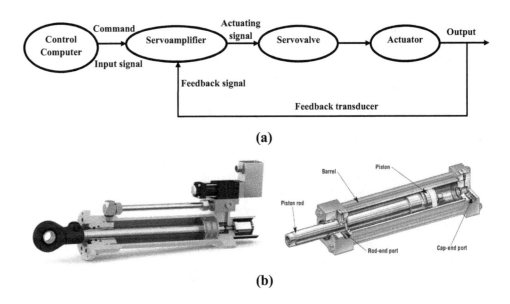

Fig. 5 a Servo-control electrohydraulic diagram; **b** schematic of a servo-hydraulic actuator (http://www.besmaklab.com/Products/133/Single-Ended_Servo_Hydraulic_Actuators/9)

(a)

(b)

All structures are modelled with the floor thicknesses and storey heights of 0.3 (m) and 3.3 (m), respectively. Actuators were used in the braces to control the structures. Using active brace control is one of the prominent strategies in control problems. This system is composed of pre-stressed systems or braces connected to the structure which are controlled by means of an electrohydraulic servocontrol system as shown in Fig. 5a. Moreover, a schematic view of a servo-hydraulic actuator is presented in Fig. 5b.

The hypotheses considered in the analysis and modelling are:

1. Damping for all the models are considered as 3 % of the critical.

2. To prevent saturation in the actuator, a constraint was defined as:

$$0.1 < \eta + \mu < 0.46, \tag{12}$$

where η and μ are the available values in the α and β matrices which are used to obtain the gain matrix. Saturation occurs in the actuators whenever the actuators work with their maximum capacity and, therefore, become unable to tolerate other loads. The main reason for using different heights for buildings is to investigate height effects on the actuator positions by considering near- and far-field earthquakes.

3. To calculate the floor weights, the Iranian national loading code was used.

Table 2 Nominated models

Name	Storey number	No of actuators	Method used for placement of actuators
M21-3-P	21	3	Optimization by using the PSO algorithm
M15-2-P	15	2	Optimization by using the PSO algorithm
M9-1-P	9	1	Optimization by using the PSO algorithm
M21-3-A	21	3	Arbitrary
M15-2-A	15	2	Arbitrary
M9-1-A	9	1	Arbitrary

The Chi–Chi and Kobe records as both near- and far-field ground motions were used to analyse models listed in Table 2 (http://peer.berkeley.edu/products/strong_ground_motion_db.html).

Properties of near- and far-field earthquakes

Obviously, there are some differences in the properties of near- and far-field earthquakes. Therefore, it seems necessary to investigate the effects of these differences on the buildings and to classify these effects. A distance shorter than 15 km from a fault line is referred to as near-fault zone; otherwise, it is known as far-field zone. In the near-field zone, earthquake effects depend on three main factors, namely (1) rupture mechanism; (2) rupture propagation directions with respect to the site, and (3) permanent displacement due to fault slippage. These factors create two phenomena, which are rupture directivity and step fling. Rupture directivity is also divided into two phenomena, which are forward directivity and backward directivity. Forward directivity effects lead to horizontal oscillations in the direction perpendicular to the fault line in the form of a horizontal pulse, which has much more significant effects on the structures in comparison with a parallel pulse to the fault line. These pulses lead to an increase in the nonlinear deformation demands of the structures. Near-fault ground motions have short duration with high amplitude and high to medium oscillation periods (International Institute of Earthquake Engineering and Seismology 2007; Alavi 2001; Galal and Ghobarah 2006; Stewart et al. 2001). The recorded databases of Kobe and Chi–Chi earthquakes were used to analyse the structures. Regard of FEMA 356, the geotechnical specifications should be taken into account in selection process of the earthquake record databases (Federal Emergency Management Agency 2000). Therefore, the frequency contents, spectrum, effective duration, and type of soil could be varied regard of construction site (Federal Emergency Management Agency 2000).

Pole Assignment controller design

In this study, Pole Assignment was used as the control method. The equation of motion of a multi-degree-of-freedom (MDOF) system by considering a force control under the effect of a specific excitation is:

$$[M]\ddot{X} + [C]\dot{X} + [K]\{X\} = -[M]\{I\}\ddot{X}_g - \{U_C\}, \quad (13)$$

where $\{Uc\}$ is the control force vector which has a dimension equal to that of the displacement vector $(n \times 1)$. The negative sign on the right hand side of Eq. (13) shows that the applied force control is in the same direction with the formed internal resistance due to damping and stiffness of the structure. $[M]$, $[C]$ and $[K]$ are the mass, damping and stiffness matrices, respectively, and have the dimensions of $(n \times n)$. In Eq. (13), $\{I\}$ is the unit vector with the dimension of $(n \times 1)$ and x_g is the earthquake acceleration record. By transforming the equation of motion into state-space, Eq. (13) can be rewritten in the following form:

$$\{\dot{q}\} = [A]\{q\} + [Be]\ddot{x}_g + [B_U]\{u_c\}, \quad (14)$$

where $[A]$ is the system matrix, $[B_u]$ the actuator position matrix, $[Be]$ the vector of external excitation position and $\{q\}$ the space vector. These matrices and vectors are given by:

$$[B_U] = \begin{bmatrix} 0 \\ -M^{-1} \end{bmatrix} \quad [Be] = \begin{Bmatrix} 0 \\ -I \end{Bmatrix},$$

$$q = \begin{Bmatrix} x \\ \dot{x} \end{Bmatrix} \quad [A] = \begin{bmatrix} 0 & I \\ -M^{-1}K & -M^{-1}C \end{bmatrix}. \quad (15)$$

The force control U_c is obtained by multiplying the gain matrix in the space vector:

$$\{U_C\} = [F]\{q\}. \quad (16)$$

The gain matrix is replaced by:

$$F = [F_K, F_C], \quad (17)$$

where F_k and F_C are the stiffness and damping type of matrices with dimensions of $(n \times n)$. These components can be obtained by the following equations:

$$[F_C] = [\alpha][C] \quad [F_K] = [\beta][K'], \tag{18}$$

in which $[\alpha]$ and $[\beta]$ are diagonal matrices defined by the following matrices:

$$[\alpha] = \begin{bmatrix} \alpha_1 & 0 & 0 & 0 & 0 \\ 0 & \alpha_2 & 0 & 0 & 0 \\ 0 & 0 & . & 0 & 0 \\ 0 & 0 & 0 & . & 0 \\ 0 & 0 & 0 & 0 & \alpha_n \end{bmatrix}$$

$$[\beta] = \begin{bmatrix} \beta_1 & 0 & 0 & 0 & 0 \\ 0 & \beta_2 & 0 & 0 & 0 \\ 0 & 0 & . & 0 & 0 \\ 0 & 0 & 0 & . & 0 \\ 0 & 0 & 0 & 0 & \beta_n \end{bmatrix}. \tag{19}$$

The stiffness and damping matrices are obtained from the stiffness and damping properties:

$$[k'] = \begin{bmatrix} k_1 & 0 & 0 & 0 & 0 \\ 0 & k_2 & 0 & 0 & 0 \\ 0 & 0 & . & 0 & 0 \\ 0 & 0 & 0 & . & 0 \\ 0 & 0 & 0 & 0 & k_n \end{bmatrix} \quad C_i = 2\xi\omega_i M_i \tag{20}$$

$$M_i = \{\Phi_i^T\}[m]\{\Phi_i\},$$

where $\{\varphi_i\}$ and ω_i are related to the mode and frequency of the ith structure, respectively. By substituting Eq. (16) into Eq. (14), the following equation will be obtained:

$$\{\dot{q}\} = ([A] + [Bu][F])\{q\} + \{Be\}\ddot{x}_g. \tag{21}$$

The new system matrix is defined as:

$$[A_{con}] = [A] + [Bu][F]. \tag{22}$$

Substituting Eq. (22) into Eq. (21) leads to the following equation:

$$\{\dot{q}\} = [A_{con}]\{q\} + \{Be\}\ddot{x}_g. \tag{23}$$

In this paper, by means of particle swarm optimization (PSO) algorithm, the optimum values of $[\alpha]$ and $[\beta]$ matrices are calculated in such a way that the obtained gain matrix modifies the system to satisfy the objective function. By performing this process, the best actuator placements are defined to control the structure in the direction of objective function. In this study, the objective function is defined as:

$$Z = 0.1Z_1 + 0.45Z_2 + 0.45Z_3, \tag{24}$$

where the components Z_1, Z_2 and Z_3 are defined as:

$$Z_3 = \min \frac{\text{Max. Controlled Force}}{\text{Max. uncontrolled Force}},$$
$$Z_2 = \min \frac{\text{Max. Controlled Drift}}{\text{Max. Uncontrolled Drift}},$$
$$Z_1 = \min \frac{\text{Max. Controlled Dis.}}{\text{Max. uncontrolled Dis.}}.$$

Using multi-objective functions lead to the optimum placement, capacity and number of actuators in comparison with the time a single objective function is used. Using energy terms in the multi-objective functions leads to improvement in the control process of structures.

Particle swarm optimization (PSO) algorithm

Particle swarm optimization algorithm is used for optimizing difficult numerical functions and, based on the metaphor of human social interaction, is capable of mimicking the ability of human societies to process knowledge (Shayeghi et al. 2009). This algorithm has roots in two main component methodologies: (1) artificial life (such as bird flocking, fish schooling and swarming); (2) evolutionary computation (Kenedy and Eberhart 1995). The main issue in this algorithm is that potential solutions are flown through hyperspace and are accelerated towards better or more optimum solutions. Particles adjust their flights based on the flying experiences of themselves and their companions. It keeps the rout of its coordinates in hyperspace which is associated with its previous best fit solution and its peer corresponding to the overall best value acquired thus far by any other particles in the population. Vectors are taken as particle presentations, since most optimization problems are convenient for such variable presentations (Shayeghi et al. 2009). Actually, the fundamental principles behind swarm intelligence are adaptability, diverse response, proximity, quality and stability. It is adaptive, based on the change of the best group value. The response assignments between the individual and group values ensure a diversity of responses. The higher-dimensional space calculations of the PSO concept are needed to be done over a series of time steps (Shayeghi et al. 2009). The population is defined as the quality factors of the previous best individual values and the previous best group values. The principle of stability and state in the PSO algorithm are functioned to the population changes and the best group value changes, respectively (Kennedy et al. 2001; Clerc and Kennedy 2002). According to (Shayeghi et al. 2008), the optimization technique can be used to solve similar problems as the GA algorithm, and not involved with the difficulties of GA problems (Shayeghi et al. 2008). By observing the obtained results from the analysed problems solved by the PSO algorithm, it was found that it was robust in solving problems featuring nonlinearity, non-differentiability and high dimensionality. The PSO algorithm is the search method to improve the speed of convergence and find the global optimum value of the fitness function (Shayeghi et al. 2009).

PSO begins with a population of random solutions "particles" in a D-dimension space. The ith particle is represented by Xi = (xi1, xi2,...,xiD) (Shayeghi et al. 2009). Each particle keeps the rout coordinates in hyperspace, associated with the fittest solution. The value of the fitness for particle ith (pbest) is also stored as Pi = (pi1, pi2,...,piD). The PSO algorithm keeps rout to approach the overall best value (gbest), and its location, obtained thus far by any particle in the population. The PSO algorithm consists of a step, involving changing the velocity of each particle towards its pbest and gbest according to Eq. (25).

The velocity of particle i is represented as Vi = (vi1, vi2... viD). Acceleration is weighted by a random term, with separate random numbers being generated for acceleration towards pbest and gbest values. Then, the ith particle position is updated based on Eq. 26 (Kennedy et al. 2001):

$$v_i(t) = \phi \, v_i(t-1) + r_1 c_1 (\vec{x}_{pbest} - \vec{x}_i) + r_2 c_2 (\vec{x}_{gbest} - \vec{x}_i),$$
(25)

$$\vec{x}_i(t) = \vec{x}_i(t-1) + \vec{v}_i(t),$$
(26)

where C_1 and C_2 are acceleration coefficients. Kenedy showed that to ensure a stabilized solution, the sum of these coefficients must be less than 4; otherwise, velocity and particle positions tend to infinity (Clerc and Kennedy 2002). ϕ represents the inertia weights for which the following equation must be satisfied:

$$\phi \rangle 0.5(C_1 + C_2) - 1.$$
(27)

A flowchart is presented in Fig. 6 which better illustrates the mechanism of this algorithm.

Analytical results

Discussions

According to near- and far-field records of Kobe and Chi–Chi earthquakes (Fig. 7), structures with 9, 15 and 21 stories were analysed in MATLAB software. The positions of the actuators and the objective-function values are listed in Tables 3, 4, 5 and 6.

To assess the accuracy and sensitivity of the results, two records were used for analysing the structures. Chi–Chi earthquake has a long duration which can have different effects on the structures with different frequencies compared to Kobe earthquake that has a short duration.

For the structures with 21 stories, three actuators were installed on the top, middle and first floors regardless of the type of the external excitation zone. According to the obtained results shown in Fig. 8, for the tall building (21 stories), 67 % of the actuators are placed at the upper half of the structure and others are placed at the lower half. Moreover, for medium-rise structures (15 stories), 75 % of the actuators are placed at the upper half of the structure, while the others are installed on the ground floors. Finally, for the short building, the actuator position was dependent on the type of the external excitation zone. As seen in Fig. 8, structures having 21, 15, and 9 stories have 1, 0.71 and 0.42 height ratios, respectively.

As indicated in Fig. 9, the maximum displacement of the controlled and uncontrolled cases at the floor level throughout the structure height highlights the effect of adopted strategy for controlling the structures.

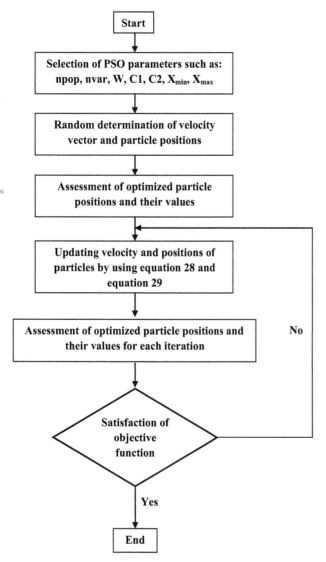

Fig. 6 Flowchart of the PSO algorithm

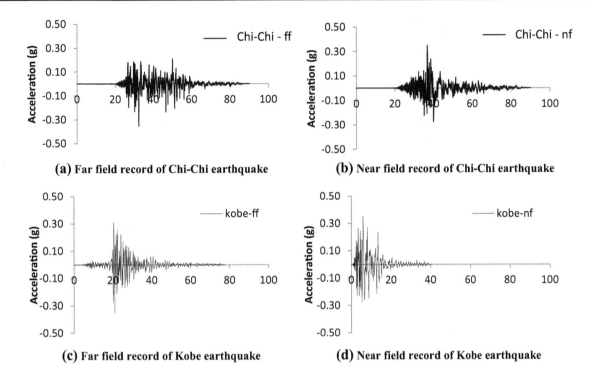

(a) Far field record of Chi-Chi earthquake

(b) Near field record of Chi-Chi earthquake

(c) Far field record of Kobe earthquake

(d) Near field record of Kobe earthquake

Fig. 7 Records of Chi–Chi and Kobe earthquake in both near and far field

Table 3 Actuator position and objective-function value for Kobe earthquake in the near field

Model	Actuator position	α	β	Objective function value (Z)
M21-3-P	1&16&18	0.25	0.2	0.1218
M15-2-P	9&13	0.25	0.2	0.1451
M9-1-P	7	0.25	0.2	0.2085
M21-3-A	7&14&21	0.2	0.25	0.1488
M15-2-A	7&15	0.2	0.25	0.1911
M9-1-A	6	0.2	0.25	0.2304

Table 4 Actuator position and objective-function value for Kobe earthquake in the far field

Model	Actuator position	α	β	Objective-function value (Z)
M21-3-P	1&16&18	0.25	0.2	0.1123
M15-2-P	1&13	0.25	0.2	0.2263
M9-1-P	8	0.25	0.2	0.2024
M21-3-A	7&14&21	0.2	0.25	0.1252
M15-2-A	7&15	0.2	0.25	0.2468
M9-1-A	6	0.2	0.25	0.2415

Table 5 Actuator position and objective-function value for Chi–Chi earthquake in the near field

Model	Actuator position	α	β	Objective-function value (Z)
M21-3-P	1&20&21	0.25	0.2	0.1101
M15-2-P	1&13	0.25	0.2	0.1804
M9-1-P	5	0.25	0.2	0.3187
M21-3-A	7&14&21	0.2	0.25	0.1257
M15-2-A	7&15	0.2	0.25	0.2043
M9-1-A	6	0.2	0.25	0.315

Table 6 Actuator position and objective-function value for Chi–Chi earthquake in the far field

Model	Actuator position	α	β	Objective-function value (Z)
M21-3-P	2&16&18	0.25	0.2	0.1158
M15-2-P	9&13	0.25	0.2	0.138
M9-1-P	9	0.25	0.2	0.2053
M21-3-A	7&14&21	0.2	0.25	0.1344
M15-2-A	7&15	0.2	0.25	0.1825
M9-1-A	6	0.2	0.25	0.2489

Fig. 8 Comparing actuator positions in near- and far-field zones by changing the height ratio

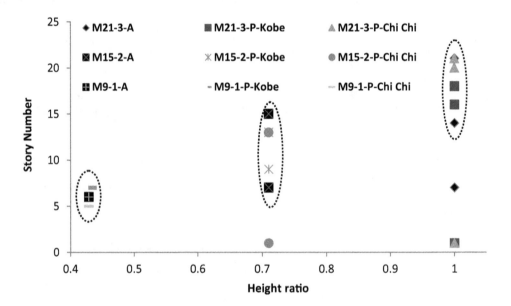

Near field earthquake zones

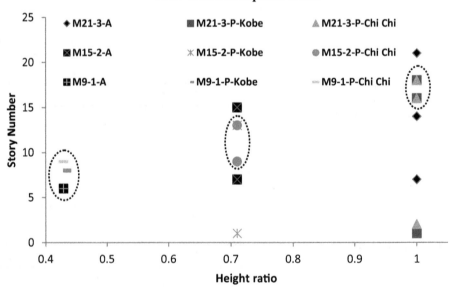

Far field earthquake zones

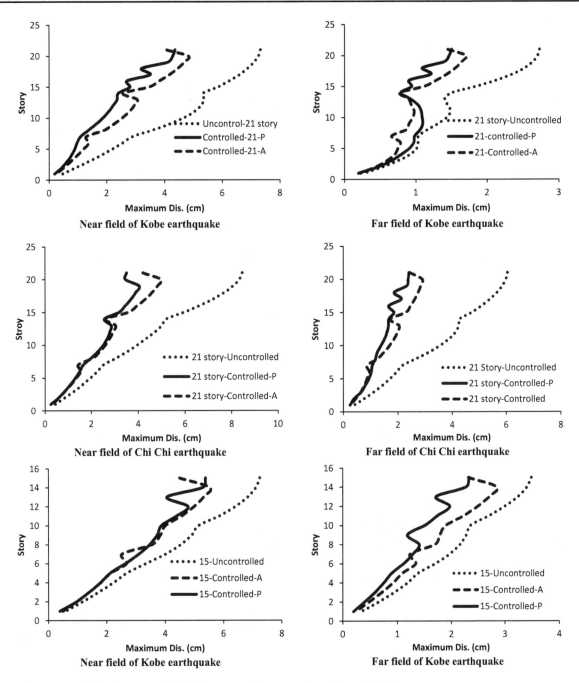

Fig. 9 Maximum controlled and uncontrolled displacements of structures throughout the height

Performance criteria

Some performance criteria introduced by Yang were used in this study to evaluate the controllers (Yang et al. 2004). The first criterion is related to the ability of the controller to reduce the maximum floor root mean square (RMS) acceleration:

$$J_1 = \max(\sigma_{\ddot{x}n})/\sigma_{\ddot{x}l}, \tag{28}$$

where $\sigma_{\ddot{x}_n}$ is the RMS acceleration of the storey in which the actuator is installed. $\sigma_{\ddot{x}_l}$ is the uncontrolled RMS

acceleration of the top storey which does not have any controller or actuator. The second criterion is the average reduction in the acceleration of the controlled floors:

$$J_2 = \frac{1}{n} \sum \frac{\sigma_{\ddot{x}_i}}{\sigma_{\ddot{x}_0}}, \tag{29}$$

where n is number of floors in which the actuators were installed. $\sigma_{\ddot{x}_i}$ and $\sigma_{\ddot{x}_0}$ represent RMS acceleration in controlled and uncontrolled stories, respectively. The third and fourth criteria are used for evaluating the top floor displacements:

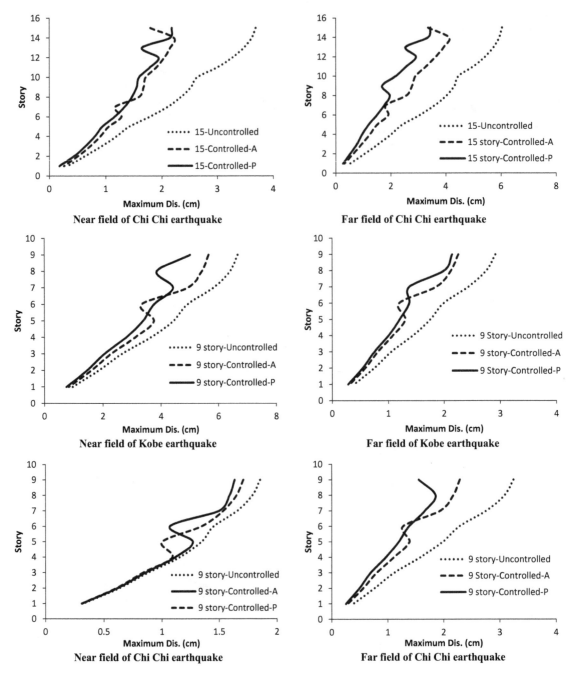

Fig. 9 continued

$$J_3 = \frac{\sigma_{x_l}}{\sigma_{x_0}}, \tag{30}$$

$$J_4 = \frac{1}{n} \sum \frac{\sigma_{x_i}}{\sigma_{x0}}, \tag{31}$$

where σ_{x_i} and σ_{x_0} are the uncontrolled and controlled RMS displacements at the top floor level, respectively. Furthermore, $\sigma_{\ddot{x}_i}$ and $\sigma_{\ddot{x}_0}$ represent the controlled and uncontrolled RMS accelerations in the storey, respectively. Three other criteria are used to assess the peak responses of the structure:

$$J_5 = \frac{\text{Max}(\ddot{x}_i)}{\ddot{x}_{l0}}, \tag{32}$$

Where $\ddot{x}_l$, $\ddot{x}_{l0}$, x_i and x_0 are the peak controlled and uncontrolled accelerations and controlled and uncontrolled displacements, respectively.

$$J_6 = \frac{1}{n} \sum \frac{\ddot{x}_i}{\ddot{x}_0}, \tag{33}$$

$$J_7 = \frac{1}{n} \sum \frac{x_i}{x_0}. \tag{34}$$

Table 7 Criteria performance for structures located in near-field zone of Kobe earthquake

Name of models	J_1	J_2	J_3	J_4	J_5	J_6	J_7
M21-3-P	0.4143	0.6171	0.5950	0.4346	0.5083	0.6501	0.4678
M21-3-A	0.9249	0.7135	0.5560	0.4946	0.5667	0.4566	0.5572
M15-2-P	0.4808	0.5803	0.7422	0.6298	0.6045	0.7179	0.7777
M15-2-A	0.5728	0.6362	0.6196	0.6238	0.7727	0.7850	0.8083
M9-1-P	0.5784	0.6133	0.7534	0.6024	0.5230	0.5570	0.7797
M9-1-A	0.4339	0.6437	0.8487	0.6752	0.4379	0.6519	0.8532

Table 8 Criteria performance for structures located in far-field zone of Kobe earthquake

Name of models	J_1	J_2	J_3	J_4	J_5	J_6	J_7
M21-3-P	0.4064	0.5737	0.5479	0.5268	0.6083	0.7306	0.7838
M21-3-A	0.8657	0.7176	0.5293	0.5589	0.6904	0.5281	0.7237
M15-2-P	0.5092	0.5944	0.6716	0.5243	0.5551	0.7762	0.6407
M15-2-A	0.6134	0.6744	0.6672	0.6484	0.8003	0.8258	0.7413
M9-1-P	0.5666	0.7088	0.7355	0.5657	0.5578	0.7324	0.7122
M9-1-A	0.4430	0.6891	0.7741	0.6003	0.4556	0.7154	0.7514

Table 9 Criteria performance for structures located in near-field zone of Chi–Chi earthquake

Name of models	J_1	J_2	J_3	J_4	J_5	J_6	J_7
M21-3-P	0.3860	0.4369	0.4119	0.4609	0.6084	0.6804	0.5569
M21-3-A	0.4785	0.5456	0.4969	0.5219	0.6621	0.6568	0.6587
M15-2-P	0.5303	0.6074	0.5913	0.5176	0.5553	0.6804	0.639
M15-2-A	0.6673	0.6316	0.4882	0.5224	0.8285	0.8985	0.7364
M9-1-P	0.5554	0.7342	0.9229	0.7425	0.3280	0.7625	0.7425
M9-1-A	0.5191	0.7185	0.8823	0.7356	0.3399	0.8263	0.9636

Table 10 Criteria performance for structures located in the far-field zone of Chi–Chi earthquake

Name of models	J_1	J_2	J_3	J_4	J_5	J_6	J_7
M21-3-P	0.4098	0.4678	0.3985	0.3878	0.4417	0.7123	0.4964
M21-3-A	0.5468	0.4958	0.4041	0.4023	0.6082	0.6780	0.5221
M15-2-P	0.3877	0.4120	0.5701	0.4153	0.5196	0.6046	0.4909
M15-2-A	0.5243	0.5087	0.5562	0.5295	0.6266	0.6215	0.6238
M9-1-P	0.5554	0.5554	0.4792	0.4792	0.5568	0.5568	0.6665
M9-1-A	0.4363	0.6119	0.7061	0.5577	0.4269	0.5798	0.7009

The smaller the numerical values for these criteria, the better is the performance of the controller. According to the obtained results listed in Tables 7, 8, 9 and 10, using this control method led to control of both acceleration and displacement in the near- and far-field zones in such a way that this reduction was more in the structures in which the optimized PSO algorithm had been used rather than the ones without this algorithm. In Fig. 10, the efficiency of using the algorithm control with respect to an arbitrary one in the defined performance criteria was investigated by varying the height and earthquake frequency.

The positive values in Fig. 10 represent lower efficiency in using the optimized PSO algorithm compared with not using it. According to the results, the J_3, J_4, J_2, J_6, J_7 criteria illustrate positive effects of using algorithm control to control short buildings for both the external excitation zones. On the other hand, for the J_1 and J_5 criteria, not using the PSO algorithm in the installed actuators led to a better performance in the defined criteria. The overall results showed that the criteria of J_3, J_4, J_6, J_7, J_2 were reduced by 20, 11.11, 5, 5.9 and 5 %, on average, respectively, for all the earthquakes in the short buildings. Furthermore, except for the J_3 criterion, the other performance criteria for the structures with medium heights have better performance in the buildings with optimized actuators compared to those with unoptimized actuators. The J_1, J_2, J_4, J_5, J_6, J_7 criteria led to average reductions of 21.28, 14.66, 15.32, 21.74, 4.99 and 14.99 %, respectively, for the

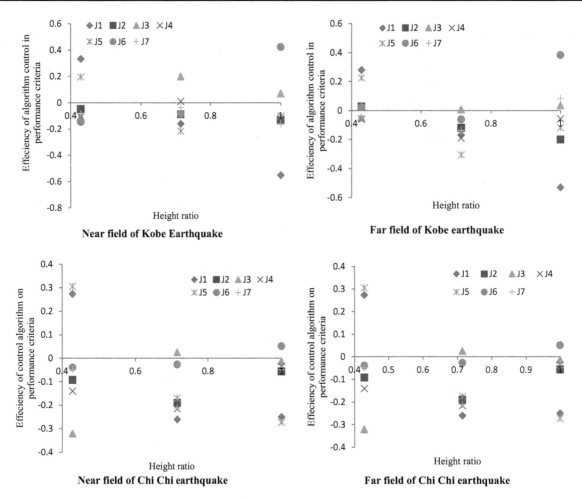

Fig. 10 Efficiency variations of defined performance criteria throughout the height ratio

structures with medium heights for both types of external excitation zones. The adopted control algorithm for the tall buildings completely depends on the type of the external excitation zone. All the introduced criteria except J_6 showed a good performance in the structure in which the control algorithm was used in comparison with the case in which no control algorithm was used. Therefore, the J_1, J_2, J_3, J_4, J_5, J_7 criteria were reduced on average by 39.59, 11.21, 1.93, 6.25, 19.23 and 4.39 %, respectively, under all applied earthquakes. By monitoring the input force to each actuator and the uncontrolled base shear, the input force for each actuator to control the buildings is determined.

According to Fig. 11, the base shear value was a function of the external excitation and the structure properties, which increased the base shear accompanied by the actuator force increase. Input actuator force was computed based on Eq. (16). Moreover, by increasing the shear force at the base level of the structure, the difference between actuator force obtained from the optimized and arbitrary cases was increased. Concerning the results obtained in

Fig. 11, the required force for actuators varied in the range of 0.4–1.2 % of the total base shear.

Conclusion

In this study, by using active actuators in the braces, three different building heights were controlled under external excitations where concrete shear walls were used as secondary lateral load-resisting system in the buildings. The placement of actuators was implemented by using optimization employing the PSO algorithm method and arbitrarily for both near- and far-field earthquake zones. Concerning the results obtained, the performance of the controller using the PSO algorithm method was significantly better than the arbitrary placement of actuators in all seismic zones. By considering the effects of seismic zone, most actuators were placed on the upper third of the structure in tall buildings, while for the moderate and short buildings, most actuators were placed on the upper half of the structure. The defined performance criteria

Fig. 11 The input energy to each actuator-uncontrolled base shear

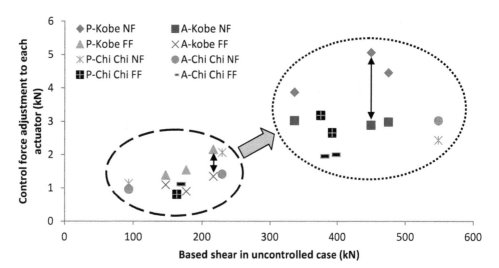

demonstrated the effectiveness of the PSO algorithm controller in both seismic zones in comparison to uncontrolled and arbitrary actuator placement. Moreover, the studies showed that the actuators used consumed much less input force to control the structure and applying actuators.

References

Alavi B, Krawinkler H (2001) Effects of near-fault ground motions on frame structures. The John A. Blume Earthquake Engineering Center, Department of Civil and Environmental engineering, Stanford University, California, Report No. 138

Amini F, Vahdani R (2008) Fuzzy optimal control of uncertain dynamic characteristics in tall building subjected to seismic excitation. J Vib Control 14:1843–1867

Clerc M, Kennedy J (2002) The particle swarm-explosion, stability, and convergence in a multidimensional complex space. IEEE Trans Evol Comput 6:58–73

Federal Emergency Management Agency (2000) Prestandard and Commentary for the Seismic Rehabilitation of Buildings (FEMA 356), prepared by the SEAOC, ATC, and CUREE Joint Venture for the Federal Emergency Management Agency, Washington, DC

Galal K, Ghobarah A (2006) Effect of near-fault earthquakes on North American nuclear design spectra. J Nuclear Eng Design 236:1928–1936

Ghali A, Neville AM, Brown TG (2003) Structural analysis: a unified classical and matrix approach. Taylor and Francis, London

International Institute of Earthquake Engineering and Seismology (IIEES) (2007) Instruction for Seismic Rehabilitation of Existing Buildings (ISREB), Management and Planning Organization of Iran, Publication

Kenedy J, Eberhart R (1995) Particle swarm optimization. In: Proceeding of the IEEE International Conference on Neural Networks, Perth, Australia. 4:1942–1948

Kennedy J, Eberhart R, Shi Y (2001) Swarm intelligence. Morgan Publishers, San Francisco

Keshavarz H, Mastali M, Gerami M, Abdollahzadeh D (2011) Investigation of performance of X-braced frames in near and far fields. In: International Conference on Earthquake Engineering and Seismology, Islam Abad, Pakistan

Kheyroddin A, Abdollahzadeh D, Mastali M (2014) Improvement of open and semi-open core wall system in tall buildings by closing of the core section in the last story. Int J Adv Struct Eng 6:1–12

Kwok NM, Ha QP, Nguyen TH, Li J, Samali B (2006) A novel hysteretic model for magnetorheological fluid dampers and parameter identification using particle swarm optimization. Sens Actuators A Phys 132:441–451

Shayeghi H, Jalili A, Shayanfar HA (2008) Multi-stage fuzzy load frequency control using PSO. J Energy Convers Manag 49:2570–2580

Shayeghi A, Shayeghi H, Eimani Kalasar H (2009) Application of PSO technique for seismic control of tall building. Int J Electr Comput Eng 4:293–300

Smith BS, Coull A (1991) Tall building structures: analysis and design. Wiley, New York

Stewart JP, Chiou SJ, Bray JD, Graves RW, Somerville PG, Abrahamson NA (2001) Ground Motion Evaluation Procedures for Performance-Based Design, A report on research conducted under grant no EEC-9701568 from the National Science Foundation. University of California, Berkeley

Symans MD, Constantinou MC (1999) Semi-active control systems for seismic protection of structures: a state-of-the-art review. J Eng Struct 21:469–487

Yang JN, Agrawal AK, Samali B, Jong-Cheng Wu (2004) Benchmark problem for response control of wind-excited tall buildings. J Eng Mech 130:437–446

3

Cyclic performance of concrete-filled steel batten built-up columns

M. S. Razzaghi[1] · M. Khalkhaliha[1] · A. Aziminejad[2]

Abstract Steel built-up batten columns are common types of columns in Iran and some other parts of the world. They are economic and have acceptable performance due to gravity loads. Although several researches have been conducted on the behavior of the batten columns under axial loads, there are few available articles about their seismic performance. Experience of the past earthquakes, particularly the 2003 Bam earthquake in Iran, revealed that these structural members are seismically vulnerable. Thus, investigation on seismic performance of steel batten columns due to seismic loads and providing a method for retrofitting them are important task in seismic-prone areas. This study aims to investigate the behavior of concrete-filled batten columns due to combined axial and lateral loads. To this end, nonlinear static analyses were performed using ANSYS software. Herein, the behaviors of the steel batten columns with and without concrete core were compared. The results of this study showed that concrete-filled steel batten columns, particularly those filled with high-strength concrete, may cause significant increases in energy absorption and capacity of the columns. Furthermore, concrete core may improve post-buckling behavior of steel batten columns.

Keywords Steel batten columns · Buckling · Numerical analysis · Lateral loads

✉ M. S. Razzaghi
razzaghi.m@gmail.com

[1] Qazvin Branch, Islamic Azad University, Qazvin, Iran

[2] Science and Research Branch, Islamic Azad University, Tehran, Iran

Introduction

During the past decades, several researches have concentrated on the performance of steel built-up batten columns due to axial loads (Hosseini Hashemi and Jafari 2009, 2012). Following the Bam earthquake of 2003 in southeastern Iran, several steel structures were damaged (Eshghi et al. 2003). Failure of the built-up batten columns was one of the most observable failure modes in damaged buildings (Hosseini Hashemi and Jafari 2004; Hosseinzadeh 2004). In most of these studies, the behavior of the batten columns due to gravitational loads was investigated. In other words, most of the researchers neglected the effect of the lateral loads on the behavior of steel batten columns (Hosseini Hashemi and Jafari 2004). Following the Bam earthquake, some studies were conducted to investigate the behavior of batten columns due to the earthquake. Razzaghi et al. (2010) investigated the performance of built-up batten columns due to axial and cyclic lateral loads. The results of this study revealed that laterally loaded batten columns had extremely unstable behaviors in the post-buckling region. It was also shown that existing code provisions are not sufficient for seismic design of batten columns. Furthermore, Hosseini Hashemi and Hassanzadeh (2008) studied the performance of the damaged steel building with batten columns due to the Bam earthquake.

Based on the seismic vulnerability of steel batten columns and the large number of buildings with these types of columns in areas with high seismic action potential, retrofit methods are required for steel batten columns. On the other hand, concrete-filled steel tubular columns are widely used in steel structures. Several numerical and experimental investigations revealed the acceptable performance of concrete-filled tubes under different loading conditions (Cai and He 2007; de Oliviera et al. 2010). Hence, it seems that filling the hollow space between column chords would be a suitable method for improvement of seismic performance of built-up batten columns.

This paper aims to investigate the performance of concrete-filled built-up batten columns (CFBBC) as both retrofitted columns and newly designed and/or constructed columns. To this end, nonlinear static analyses were performed using the ANSYS (SAS 2010) software. Herein, parametric studies on the effects of compressive strength of concrete, distances of batten plates and axial loads on the performance of CFBBC were conducted.

Modeling

To conduct a parametric study on the performance of axially and laterally loaded CFBBC, the finite element method was used. In all of the models, both material and geometric nonlinearities were considered.

Geometric specifications of models

Geometric specifications of steel batten columns were adopted from the columns of the ground floor of a building damaged in the Bam earthquake as reported by Hosseini Hashemi and Hassanzadeh (2008). The target building was of steel frame with an X-bracing system. The columns had two IPE 160 chords, 17 cm apart, and battened by 18 × 10 × 1 cm batten plates at a distance of 50 cm c/c. The height of the ground floor was 5.2 m. The geometric specifications of the models are summarized in Table 1. As indicated in Table 1, there are two main categories of models: columns without concrete cores and columns with concrete cores. It should be noted that for those column which were originally designed and fabricated as CFBBC, the axial loads would be distributed to concrete and steel according to their stiffness. Such distributions may not occur in retrofitted batten columns. Hence as indicated in the second column of Table 1, different distribution of axial loads are considered in CFBBC models.

Elements used for FE analyses

Three-dimensional solid elements (Solid 65) have been used to model the concrete core. This element is defined by eight nodes having three translational degrees of freedom at each node. This solid element is capable of cracking in tension and crushing in compression, plastic deformation and creep (SAS 2010). Three-dimensional solid elements (Solid 45) were used to model the steel batten columns. This element is defined by eight nodes having three translational degrees of freedom at each node. This element has suitable compatibility with the solid element of the concrete core. The element has plasticity, stress stiffening, large deflection and large strain capabilities (SAS 2010). Three-dimensional node to node contact element (Contact 178) was used to model the contact between the steel wall and concrete core. The element is located between two adjacent nodes of steel wall and the concrete core and is capable of modeling the separation, sliding and contact between two nodes during the loading process. The element has two nodes with three translational degrees of freedom at each node. The element is capable of supporting compression in the contact normal direction and Coulomb friction in the tangential direction (SAS 2010).

Table 1 Specifications of models

Compressive strength of concrete (MPa)	Type of column	Distance of batten plates (mm)	Distribution of axial loads	Model name
–	Hollow	500	Steel section	B-1
–	Hollow	250	Steel section	B-2
–	Hollow	250 and 500[a]	Steel section	B-3
20	Concrete filled	500	Steel section	C-1-S-200
20	Concrete filled	250	Steel section	C-2-S-200
20	Concrete filled	250 and 500[a]	Steel section	C-3-S-200
60	Concrete filled	500	Steel section	C-1-S-600
20	Concrete filled	500	Entire section according to stiffness	C-1-E-200
20	Concrete-filled	250	Entire section according to stiffness	C-2-E-200
20	Concrete filled	250 and 500[a]	Entire section according to stiffness	C-3-E-200
60	Concrete filled	500	Entire section according to stiffness	C-1-E-600
60	Concrete filled	500	10 % concrete section and 90 % steel section	C-1-E10-600
60	Concrete filled	500	30 % concrete section and 70 % steel section	C-1-E30-600

[a] The distance between the first three battens from the bottom is 250 mm and the other batten plates placed at each 500 mm

Material properties

The stress–strain behavior materials of concrete core and steel sections are shown in Fig. 1a, b, respectively. It should be noted that the multilinear isotropic stress–strain curve for ordinary concrete is computed with the equation proposed by MacGregor (1992), and for high-strength concrete is computed with the equation proposed by Popovics (1973). The elastic modulus of concrete for ordinary concrete and high-strength concrete is defined by Eqs. 1 and 2, respectively:

$$E_c = 5000\sqrt{f_c'} \quad (Mpa), \tag{1}$$

$$E_c = \left(3320\sqrt{f_c'} + 6900\right)\left(\frac{\rho}{2300}\right)^{1.5} \quad (Mpa), \tag{2}$$

in which f_c' is the compressive strength of concrete (Mpa) and ρ is the density of concrete (kg/m³).

Material nonlinearity of steel was accounted for the analyses based on Von Mises yield criterion. The kinematic hardening rule was used to define the material property in these elements. The Willam and Warnke (1974) model was used to define the failure of concrete. The concrete failure criterion can be expressed as follows:

$$\frac{F}{f_c'} - S \geq 0, \tag{3}$$

where S is the failure surface which is expressed by five input parameters: ultimate uniaxial compressive and tensile strength, ultimate biaxial compressive strength, ultimate compressive strength for a state of biaxial compression superimposed on hydrostatic stress state and ultimate compressive strength for a state of uniaxial compression superimposed on hydrostatic stress state. The failure surface in principal stress space σ_{zp} close to zero is indicated in Fig. 2.

Boundary conditions and loading pattern

The boundary conditions and loading pattern of the models are illustrated in Fig. 3. Two loading conditions were considered: constant axial loads along with monotonic increasing lateral displacement and constant axial load

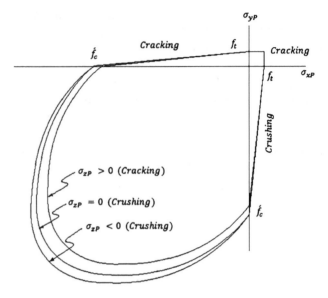

Fig. 2 Failure surface in the principal stress space close to zero (SAS 2010)

along with cyclic lateral displacement. To this end, a constant axial force of 350 kN was applied to all of the models. Coincidentally, lateral cyclic displacement (δ) is applied to the top of the columns, as indicated in Fig. 3. It is worth mentioning that the parameter (δ_y) in Fig. 3 is the lateral displacement of the yielding point of batten columns without concrete core. It should be noted that two independent conditions were considered for lateral loads: parallel to batten plates and perpendicular to them.

Results

Figure 4 indicates the pushover curves of the steel batten columns and CFBBC models. As indicated in this figure, the behavior of CFBBCs in which the axial load had just been applied to the steel chords was not acceptable. In such columns, both pre- and post-buckling performance was the same as those of built-up batten columns. Figure 4 also illustrates that even hollow batten columns with short distances of batten plates perform slightly better than

Fig. 1 The stress–strain relationship for the material

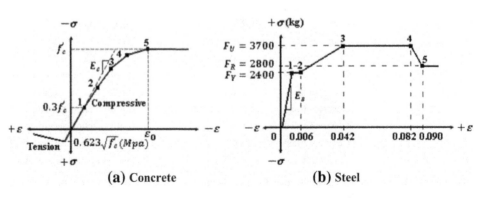

(a) Concrete   **(b) Steel**

Fig. 3 Cyclic lateral loading history and loading conditions used in FEM

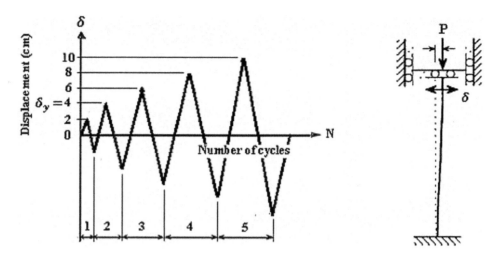

Fig. 4 Comparison of shear force–displacement curves in batten columns and CFBBCs

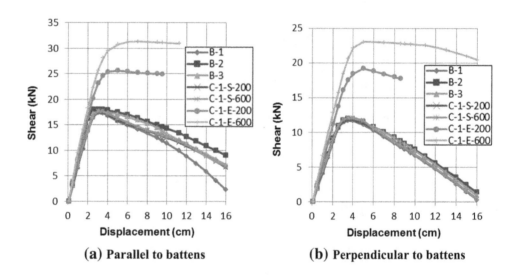

(a) Parallel to battens   **(b) Perpendicular to battens**

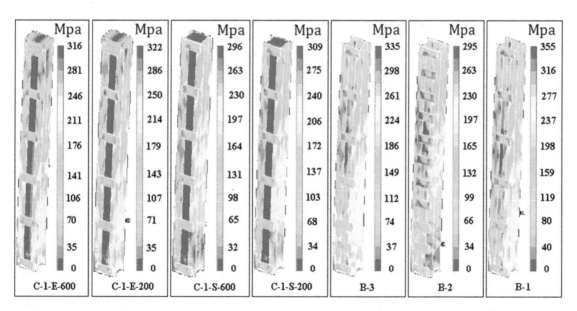

Fig. 5 Deformed shapes and Von Mises stress contours in the displacement to 9.5 cm and *parallel* battens

Fig. 6 Shear force–displacement *curves* for different CFBBCs

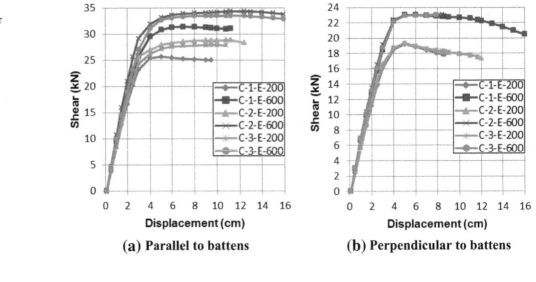

(a) Parallel to battens   (b) Perpendicular to battens

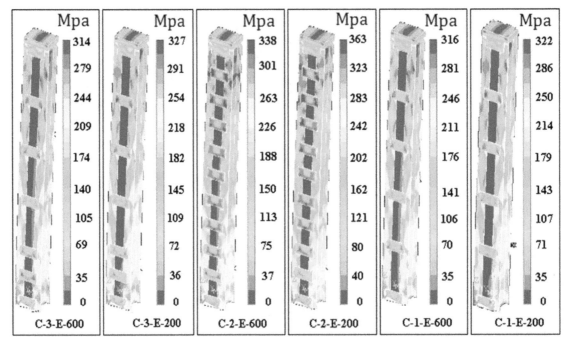

Fig. 7 Deformed shapes and Von Mises stress contours in the displacement to 9.5 cm and *parallel* battens

CFBBCs without axial load bearing of concrete core. It is obvious that in such a case, using high-strength concrete does not noticeably change the performance of CFBBC. Those CFBBCs in which the concrete core contributed to bearing axial loads, the behavior changed dramatically. Distribution of the Von Mises stress in column chords and batten plates is indicated in Fig. 5.

Comparison of shear force–displacement behavior of CFBBCs and distribution of Von Mises stress on their deformed shapes are shown in Figs. 6 and 7, respectively. As illustrated in Fig. 6a, in CFBBC columns in which concrete cores contributed in bearing the axial load

according to their stiffness, the buckling was not the major failure mode of the columns. Minor local buckling happened in some of the models after formation of plastic hinges. In locally buckled columns, the post-buckling behavior was rather stable. Concrete-filled batten columns with high-strength concrete cores had the highest ultimate strength. Variation of the compressive strength of concrete from 20 to 60 MPa made 13–20 % change in the ultimate strength of CFBBCs. Furthermore, decreasing the spacing of batten plates at the bottom of the column increases the ultimate capacity of the columns by about 5 %. As indicated in Fig. 6b when the lateral loads are applied to the

Fig. 8 The envelope curves of hysteretic loops

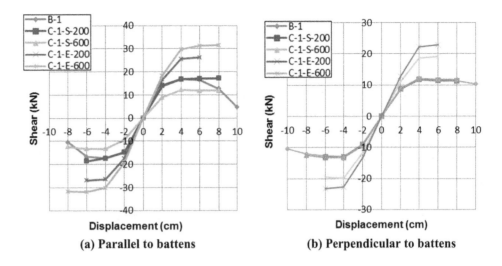

(a) Parallel to battens   (b) Perpendicular to battens

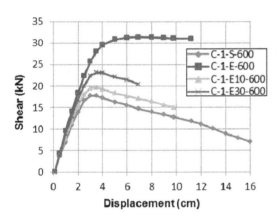

Fig. 9 Shear force–displacement *curves, parallel* to battens

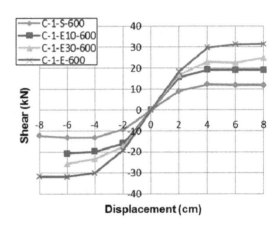

Fig. 10 The envelope curves of hysteretic loops, *parallel* to battens

column in a direction perpendicular to the batten plates, the spacing of batten plates do not have any contribution in the bearing capacity of CFBBCs.

Figure 8 shows the envelope curves of hysteretic loops for the batten columns with and without concrete cores. As illustrated in this figure, there is an observable load drop in envelope curve of the steel batten column. Such a drop happened because of buckling. It is observable in Fig. 8a that no drop happened in the envelope curve of the hysteresis loops of CFBBCs. In other words, even in CFBBCs without load carrying in the concrete core, the hysteretic behavior is noticeably more stable than that of steel batten columns. Figure 8b shows that load-carrying cores had observable change in the hysteretic behavior of steel batten columns.

To investigate the effects of load carrying of the concrete core on the performance of the CFBBCs, the pushover curves of CFBBCs with different axial load carrying in concrete cores are compared in Fig. 9. As indicated in this figure, the ultimate strengths of the CFBBCs were noticeably increased by increasing the axial load carrying

of the concrete core. Furthermore, similar results were observable in the hysteretic behavior of these columns (See Fig. 10).

Conclusions

In this paper, parametric study on nonlinear performance of steel batten columns with and without concrete cores was conducted. To this end, nonlinear static analyses were carried out. Results of this study revealed that:

- Axial load-carrying concrete cores can increase the ultimate strength of the steel batten columns.
- Despite the steel batten columns in CFBBCs with axial load-carrying cores, buckling is not the major failure mode of the column.
- In CFBBCs with axial load-carrying cores, changing the compressive strength of concrete from 20 to 60 Mpa causes about 13–20 % increase in the ultimate capacity of the column.

- The spacing of the bottom batten plates in the built-up batten columns with and without concrete core makes a noticeable contribution in ultimate capacity and hysteretic behaviors of columns.

Based on the results of this study, using concrete core would be a suitable approach for strengthening of the built-up batten columns. But in implementation of this method, attention should be given to ensure that axial loads are distributed between both steel chords and concrete cores.

Acknowledgments The authors would like to appreciate Mr. Emamisaleh of QIAU for his valuable support.

References

Cai J, He Z (2007) Eccentric-loaded behavior of square CFT columns with binding bars. J Build Struct 4:004

de Oliveira WLA, De Nardin S, de Cresce El ALH, El Debs MK (2010) Evaluation of passive confinement in CFT columns. J Constr Steel Res 66(4):487–495

Eshghi S, Zare M, Asadi K, Razzaghi M, Ahari M, Motamedi M (2003) Reconnaissance report on 26 December 2003 Bam earthquake. International Institute of Earthquake Engineering and Seismology (IIEES), Report in Persian

Hosseini Hashemi B, Jafari MA (2009) Experimental evaluation of elastic critical load in batten columns. J Constr Steel Res 65(1):125–131

Hosseini Hashemi B, Jafari MA (2012) Evaluation of Ayrton-Perry formula to predict the compressive strength of batten columns. J Constr Steel Res 68(1):89–96

HosseiniHashemi B, Hassanzadeh M (2008) "Study of a semi-rigid steel braced building damaged in the Bam earthquake". J Constr Steel Res 64:704–721

HosseiniHashemi B, Jafari MA (2004) "Performance of batten columns in steel buildings during the Bam Earthquake of 26 December 2003". J Seismol Earthq Eng 5(4):101–110

Hosseinzadeh NA (2004) Lessons learned from steel braced buildings damaged by the Bam Earthquake of 26 December 2003. JSEE J Seismol Earthq Eng 5(4):111–121

MacGregor JG (1992) Reinforced concrete mechanics and design. Prentice-Hall Inc, Englewood Cliffs

Popovics S (1973) A numerical approach to the complete stres—strain curve of concrete. Cem Concr Res 3(5):553–599

Razzaghi MS, Jafari MA, Ahmadpour HA (2010) "Seismic performance of steel build-up batten columns". In: Proceedings of the 9th U.S. National and 10th Canadian Conference on Earthquake Engineering, Paper No.1327, Toronto, Canada

SAS (2010) ANSYS 12.0.1 Finite Element Analysis System, SAS IP, Inc

Willam KJ, Warnke EP (1974) "Constitutive model for triaxial behaviour of concrete," Seminar on Concrete Structures Subjected to Triaxial Stresses, International Association of Bridge and Structural Engineering Conference, Bergamo, Italy, pp 174

4

The effect of multi-directional nanocomposite materials on the vibrational response of thick shell panels with finite length and rested on two-parameter elastic foundations

Vahid Tahouneh[1] · Mohammad Hasan Naei[2]

Abstract The main purpose of this paper is to investigate the effect of bidirectional continuously graded nanocomposite materials on free vibration of thick shell panels rested on elastic foundations. The elastic foundation is considered as a Pasternak model after adding a shear layer to the Winkler model. The panels reinforced by randomly oriented straight single-walled carbon nanotubes are considered. The volume fractions of SWCNTs are assumed to be graded not only in the radial direction, but also in axial direction of the curved panel. This study presents a 2-D six-parameter power-law distribution for CNTs volume fraction of 2-D continuously graded nanocomposite that gives designers a powerful tool for flexible designing of structures under multi-functional requirements. The benefit of using generalized power-law distribution is to illustrate and present useful results arising from symmetric, asymmetric and classic profiles. The material properties are determined in terms of local volume fractions and material properties by Mori–Tanaka scheme. The 2-D differential quadrature method as an efficient numerical tool is used to discretize governing equations and to implement boundary conditions. The fast rate of convergence of the method is shown and results are compared against existing results in literature. Some new results for natural frequencies of the shell are prepared, which include the effects of elastic coefficients of foundation, boundary conditions, material and geometrical parameters. The interesting results indicate that a graded nanocomposite volume fraction in two directions has a higher capability to reduce the natural frequency than conventional 1-D functionally graded nanocomposite materials.

Keywords Thick shell panels · Randomly oriented straight single-walled CNTs · Two-parameter elastic foundations · Vibration analysis of structures

Introduction

Layered composite materials, due to their thermal and mechanical merits compared to single-composed materials, have been widely used for a variety of engineering applications. However, owing to the sharp discontinuity in the material properties at interfaces between two different layers, there may exist stress concentrations causing severe material failure (Weissenbek et al. 1997). Functionally graded materials are heterogeneous composite materials, in which the material properties vary continuously from one interface to the other. The advantage of using these materials is that they can survive in high thermal gradient environment, while maintaining their structural integrity. Typically, an FGM is made of a ceramic and a metal for the purpose of thermal protection against large temperature gradients. The ceramic material provides a high-temperature resistance due to its low thermal conductivity, while the ductile metal constituent prevents fracture due to its greater toughness. FGMs are now developed for general use as structural elements in extremely high temperature environments. A listing of different applications can be found in Forum (1991). Most of the studies on FGMs have

✉ Vahid Tahouneh
vahid.tahouneh@ut.ac.ir; vahid.th1982@gmail.com

Mohammad Hasan Naei
mhnaei@ut.ac.ir

[1] Young Researchers and Elite Club, Islamshahr Branch, Islamic Azad University, Islamshahr, Iran

[2] School of Mechanical Engineering, Tehran University, Tehran, Iran

been restricted to thermal stress analysis, thermal buckling, fracture mechanics and optimization (Cho and Tinsley oden 2000; Chunyu et al. 2001; Lanhe 2004). Some researches (Loy et al. 1999; Pradhan et al. 2000; Han et al. 2001; Ng et al. 2001) are based on the classical shell theory, i.e., neglecting the effect of transverse shear deformation. The application of this theory to moderately thick or thick shell structures can lead to serious errors. Using the first and higher order shear deformation theories, some modifications are done to include the effects of transverse shear deformation. In this study, the problem formulations were based on the higher order shear deformation shell theories. Yang and Shen (2003) proposed a semi-analytical approach based on Reddy's higher order shear deformation shell theory, for free vibration and dynamic instability of FGM cylindrical panels under combined static and periodic axial forces and thermal loads. Free vibration and stability of functionally graded shallow shells according to a 2-D higher order deformation theory were investigated by Matsunaga (2008). Free vibration analysis of functionally graded curved panels was carried out using a higher order formulation by Pradyumna and Bandyopadhyay (2008). They used a C0 finite element formulation to carry out the analysis. Using a 2-D higher order shear deformation theory, vibration and buckling analyses of simply supported circular cylindrical shells made of functionally graded materials (FGMs) were studied by Matsunaga (2009). He used the method of power series expansion of continuous displacement components to solve the problem. In all of the above studies the variation of the radius through the thickness was not considered and the problem formulations were based on the constant mean radius of curvature.

Two-dimensional theories reduce the dimensions of problems from three to two by introducing some assumptions in mathematical modeling leading to simpler expressions and derivation of solutions. However, these simplifications inherently bring errors and therefore may lead to unreliable results for relatively thick panels. As a result, three-dimensional analysis of panels not only provides realistic results, but also allows further physical insights, which cannot otherwise be predicted by the two-dimensional analysis. There are some studies on free vibration analysis of isotropic and composite panels and shells based on the three-dimensional elasticity formulation (Chern and Chao 2000).

Structures resting on elastic foundations with different shapes, sizes, and thickness variations and boundary conditions have been the subject of investigations, and those play an important role in aerospace, marine, civil, mechanical, electronic and nuclear engineering problems. For example, plates and shells are used in various kinds of industrial applications such as the analysis of reinforced concrete pavement of roads, airport runways and foundations of buildings. The Pasternak model (also referred to as the two-parameter model) was widely used to describe the mechanical behavior of the foundation, in which the well-known Winkler model is a special case.

The most serious deficiency of the Winkler foundation model is to have no interaction between the springs. In other words, the springs in this model are assumed to be independent and unconnected. The Winkler foundation model is fairly improved by adopting the Pasternak foundation model, a two-parameter model, in which the shear stiffness of the foundation is considered. The evident importance in practical applications, investigations on the dynamic characteristics of FGM plates and panels on elastic foundations are still limited in number. Yas and Tahouneh (2012) investigated the free vibration analysis of thick FG annular plates on elastic foundations via differential quadrature method based on the three-dimensional elasticity theory and Tahouneh and Yas (2012) investigated the free vibration analysis of thick FG annular sector plates on Pasternak elastic foundations using DQM. Tahouneh et al. (2013) studied free vibration characteristics of annular continuous grading fiber reinforced (CGFR) plates resting on elastic foundations using DQM. More recently, (Tahouneh and Naei 2014) achieved the natural frequencies of thick multi-directional functionally graded rectangular plates resting on a two-parameter elastic foundation via 2-D differential quadrature method, The proposed rectangular plates had two opposite edges simply supported, while all possible combinations of free, simply supported and clamped boundary conditions were applied to the other two edges. Farid et al. (2010) presented free vibration analysis of initially stressed thick simply supported functionally graded curved panel resting on two-parameter elastic foundation (Pasternak model), subjected in thermal environment was studied using the three-dimensional elasticity formulation. Tahouneh (2014) investigated free vibration analysis of continuous grading fiber reinforced (CGFR) FG annular sector plates on two-parameter elastic foundations under various boundary conditions, based on the three-dimensional theory of elasticity. The plates with simply supported radial edges and arbitrary boundary conditions on their circular edges were considered.

Recently, nanocomposites have significant importance for engineering applications that require high levels of structural performance and multi-functionality. Carbon nanotubes (CNTs) have demonstrated exceptional mechanical, thermal and electrical properties. These materials are considered as one of the most promising reinforcement materials for high performance structural and multi-functional composites with vast application potentials (Esawi and Farag 2007; Thostenson et al. 2001). Most studies on carbon nanotube-reinforced composites

(CNTRCs) have focused on their material properties (Esawi and Farag 2007; Thostenson et al. 2001; Dai 2002; Kang et al. 2006; Lau et al. 2006). Gojny et al. (2005) focused on the evaluation of the different types of the CNTs applied, their influence on the mechanical properties of epoxy-based nanocomposites and the relevance of surface fictionalization. Fidelus et al. (2005) investigated thermo-mechanical properties of epoxy-based nanocomposites based on low weight fraction of randomly oriented single-and multi-walled CNTs. Han and Elliott (2007) determined the elastic modulus of composite structures under CNTs reinforcement by molecular dynamic simulation and investigated the effect of volume fraction of SWNTs on mechanical properties of nanocomposites. Manchado et al. (2005) blended small amounts of arc-SWNT into isotactic polypropylene and observed the modulus increase from 0.85 to 1.19 GPa at 0.75 wt%. In addition, the strength increased from 31 to 36 MPa by 0.5 wt%. Both properties were observed to fall off at higher loading levels. These investigations and (Mokashi et al. 2007; Zhu et al. 2007) have shown that the addition of small amount of carbon nanotube in the matrix can considerably improve the mechanical, electrical and thermal properties of polymeric composites. This behavior, combined with their low density makes them suitable for transport industries, especially for aeronautic and aerospace applications where the reduction of weight is crucial in order to reduce the fuel consumption.

The properties of the CNT-reinforced composites (CNTRCs) depend on a variety of parameters including CNT geometry and the inter-phase between the matrix and CNT. Interfacial bonding in the inter-phase region between embedded CNT and its surrounding polymer is a crucial issue for the load transferring and reinforcement phenomena Shokrieh and Rafiee (2010). The traditional approach to fabricating nanocomposites implies that the nanotube is distributed either uniformly or randomly such that the resulting mechanical, thermal, or physical properties do not vary spatially at the macroscopic level. Experimental and numerical studies concerning CNTRCs have shown that distributing CNTs uniformly as the reinforcements in the matrix can achieve moderate improvement of the mechanical properties only (Seidel and Lagoudas 2006). This is mainly due to the weak interface between the CNTs and the matrix where a significant material property mismatch exists. The concept of FGM can be utilized for the management of a material's microstructure, so that the vibrational behavior of a plate/shell structure reinforced by CNTs can be improved. According to a comprehensive survey of literature, the authors found that there are few research studies on the mechanical behavior of functionally graded CNTRC structures. For the first time, Shen (2009) suggested that the nonlinear bending behavior can be

considerably improved through the use of a functionally graded distribution of CNTs in the matrix. He introduced the CNT efficiency parameter to account load transfer between the nanotube and polymeric phases.

Due to intrinsic complexity of the formulations based on the three-dimensional elasticity, powerful numerical methods are needed to solve the governing equations. The differential quadrature method (DQM) is a relatively new numerical technique in structural analysis. A review of the early developments in the differential quadrature method can be found in papers by (Bert and Malik 1997).

This paper is motivated by the lack of studies in the technical literature concerning to the three-dimensional vibration analysis of thick bidirectional nanocomposite curved panels resting on a two-parameter elastic foundation reinforced by randomly oriented straight single-walled carbon nanotubes CNTs. To the authors' best knowledge, research on the vibration of thick curved panels reinforced by randomly oriented straight single-walled carbon nanotubes which are graded in both direction including axial and radial directions has not been seen until now. The volume fractions of randomly straight single-walled carbon nanotubes SWCNTs are assumed to be graded in the thickness and also axial directions of the curved panels. The direct application of CNT properties in micromechanics models for predicting material properties of the nanotube/polymer composite is inappropriate without taking into account the effects associated with the significant size difference between a nanotube and a typical carbon fiber (Odegard et al. 2003). In other words, continuum micromechanics equations cannot capture the scale difference between the nano and micro-levels. In order to overcome this limitation, a virtual equivalent fiber consisting of nanotube and its inter-phase which is perfectly bonded to surrounding resin is applied.

This study presents a novel 2-D six-parameter power-law distribution for CNTs volume fraction of 2-D functionally graded nanocomposite materials that gives designers a powerful tool for flexible designing of structures under multi-functional requirements. Various material profiles along the radial and axial directions are illustrated by using the 2-D power-law distribution. The effective material properties at a point are determined in terms of the local volume fractions and the material properties by the Mori–Tanaka scheme. A sensitivity analysis is performed, and the natural frequencies are calculated for different sets of boundary conditions and different combinations of the geometric, material, and foundation parameters. Therefore, very complex combinations of the material properties, boundary conditions, and foundation stiffness are considered in the present semi-analytical solution approach.

Problem description

In this section, a virtual equivalent fiber consisting of a nanotube and its inter-phase which is perfectly bonded to surrounding resin is introduced to obtain the mechanical properties of the carbon nanotube/polymer composite by using the results of multi-scale FEM Shokrieh and Rafiee (2010). The equivalent fiber for SWCNT with chiral index of (10, 10) is a solid cylinder with diameter of 1.424 nm. The inverse of the rule of mixture is used to calculate material properties of equivalent fiber (Tsai et al. 2003):

$$
\begin{aligned}
E_{\text{LEF}} &= \frac{E_{\text{LC}}}{V_{\text{EF}}} - \frac{E_M V_M}{V_{\text{EF}}}, \\
\frac{1}{E_{\text{TEF}}} &= \frac{1}{E_{\text{TC}} V_{\text{EF}}} - \frac{V_M}{E_M V_{\text{EF}}}, \\
\frac{1}{G_{\text{EF}}} &= \frac{1}{G_C V_{\text{EF}}} - \frac{V_M}{G_M V_{\text{EF}}}, \\
\upsilon_{\text{EF}} &= \frac{\upsilon_C}{V_{\text{EF}}} - \frac{\upsilon_M V_M}{V_{\text{EF}}},
\end{aligned}
\tag{1}
$$

where E_{LEF}, E_{TEF}, G_{EF}, υ_{EF}, E_{LC}, E_{TC}, G_C, υ_C, E_M, G_M, υ_M, V_{EF} and V_M are longitudinal modulus of equivalent fiber, transverse modulus of equivalent fiber, shear modulus of equivalent fiber, Poisson's ratio of equivalent fiber, longitudinal modulus of composites, transverse modulus of composites, shear modulus of composites, Poisson's ratio of composites, modulus of matrix, shear modulus of matrix, Poisson's ratio of matrix, volume fraction of the equivalent fiber and volume fraction of the matrix, respectively. E_{LC}, G_C, and E_{TC} are obtained from multi-scale FEM or molecular dynamics (MD) simulations. It should be mentioned that the volume fraction of the equivalent fiber is assumed to be 7.5 % (Shokrieh and Rafiee 2010) and Poly {(mphenylenevinylene)-co-[(2,5-dioctoxy-p-phenyle) vinylene]}, referred to as (PmPV), is selected as a matrix material:

$$E^m = 2.1\,\text{Gpa}, \quad \rho^m = 1150\,\text{kg/m}^3, \quad \upsilon^m = 0.34.$$

The material properties adopted for equivalent fiber are (Shokrieh and Rafiee 2010):

$$
\begin{aligned}
E_1^{cn} &= 649.12\,\text{Gpa}, \\
E_1^{cn} &= 11.27\,\text{Gpa}, \\
\upsilon &= 0.284, \\
G^{cn} &= 5.13\,\text{Gpa}, \\
\rho^{cn} &= 1400\,\text{kg/m}^3
\end{aligned}
$$

Composites reinforced with aligned, straight CNTs

Following the standard MT derivation, one can develop the expression for effective composite stiffness C. This is obtained by using an equivalent fiber having the effective

CNT properties in the MT approach which is given as (Shi et al. 2004):

$$C = C_m + f_r \langle (C_r - C_m) A_r \rangle (f_m I + f_r \langle A_r \rangle)^{-1}, \tag{2}$$

where f_r and f_m are the fiber and matrix volume fractions, respectively. C_m is the stiffness tensor of the matrix material; C_r is the stiffness tensor of the equivalent fiber; I is the forth order identity tensor and A_r is the dilute strain-concentration tensor of the rth phase for the fiber which is given as:

$$A_r = \left[I + S(C_m)^{-1}(C_r - C_m) \right]^{-1}, \tag{3}$$

where S is Eshelby's tensor, as given by (Eshelby 1957) and (Mura 1982). The terms enclosed by angle brackets in Eq. (2) represent the average value of the term over all orientations defined by transformation from the local fiber coordinates (O-$x'_1 x'_2 x'_3$) to the global coordinates (O-$x_1 x_2 x_3$) (Fig. 1). Assume axis x_2 as the direction along the aligned nanotube. The elastic properties of the nanocomposite are determined from the average strain obtained in the representative volume element. The matrix is assumed to be elastic and isotropic, with Young's modulus E_m and Poisson's ratio υ_m. Each straight CNT is modeled as a long fiber with transversely isotropic elastic properties and has a stiffness matrix given by Eq. (1). Therefore, the composite is also transversely isotropic, with five independent elastic constants. The substitution of nonvanishing components of the Eshelby tensor S for a straight, long fiber along the x_2-direction (Shi et al. 2004) in Eq. (3) gives the dilute mechanical strain concentration tensor. Then, the substitution of Eq. (3) into Eq. (2) gives the tensor of effective elastic moduli of the composite reinforced by aligned, straight CNTs. The axial and transverse Young's modulus of the composite can be calculated from the Hill's elastic modulus as (Shi et al. 2004):

$$E_1 = n - \frac{l^2}{k}, \quad E_2 = \frac{4m(kn - l^2)}{kn - l^2 + mn}, \tag{4}$$

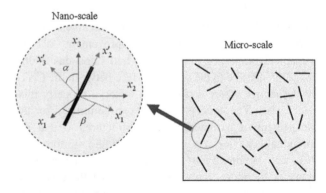

Fig. 1 Representative volume element (RVE) with randomly oriented, straight CNT

where k, l, m and n are its plane-strain bulk modulus normal to the fiber direction, cross-modulus, transverse shear modulus, axial modulus and axial shear modulus, respectively, and can be found in the Appendix. As mentioned before, the CNTs are transversely isotropic and have a stiffness matrix given below:

$$C_r = \begin{bmatrix} \dfrac{1}{E_L} & \dfrac{-\upsilon_{TL}}{E_T} & \dfrac{-\upsilon_{ZL}}{E_Z} & 0 & 0 & 0 \\ \dfrac{-\upsilon_{LT}}{E_L} & \dfrac{1}{E_T} & \dfrac{-\upsilon_{ZT}}{E_Z} & 0 & 0 & 0 \\ \dfrac{-\upsilon_{LZ}}{E_L} & \dfrac{-\upsilon_{TZ}}{E_T} & \dfrac{1}{E_Z} & 0 & 0 & 0 \\ 0 & 0 & 0 & \dfrac{1}{G_{TZ}} & 0 & 0 \\ 0 & 0 & 0 & 0 & \dfrac{1}{G_{ZL}} & 0 \\ 0 & 0 & 0 & 0 & 0 & \dfrac{1}{G_{LT}} \end{bmatrix} \qquad (5)$$

where E_L, E_T, E_Z, G_{TZ}, G_{ZL}, G_{LT}, υ_{LT}, υ_{LZ}, υ_{TZ} are material properties of the equivalent fiber which can be determined from the inverse of the rule of mixture.

Composites reinforced with randomly oriented, straight CNTs

The effective properties of composites with randomly oriented non-clustered CNTs, such as in Fig. 1, are studied in this section. The resulting effective properties for the randomly oriented CNT composite are isotropic, despite the CNTs having transversely isotropic effective properties. The orientation of a straight CNT is characterized by two Euler angles α and β, as shown in Fig. 1. When CNTs are completely randomly oriented in the matrix, the composite is then isotropic, and its bulk modulus k and shear modulus G are derived as:

$$k = k_m + \frac{f_r(\delta_r - 3K_m\alpha_r)}{3(f_m + f_r\alpha_r)}, G = G_m + \frac{f_r(\eta_r - 2G_m\beta_r)}{2(f_m + f_r\beta_r)}, \qquad (6)$$

where $\alpha_r, \beta_r, \delta_r$ and η_r can be found in the Appendix. The effective Young's modulus E and Poisson's ratio υ of the composite is given by:

$$E = \frac{9KG}{3K + G}, \quad \upsilon = \frac{3K - 2G}{6K + 2G} \qquad (7)$$

Functionally graded carbon nanotube-reinforced

Consider a bidirectional nanocomposite curved panel rested on two-parameter elastic foundations as shown in Fig. 2. A cylindrical coordinate system (r, θ, z) is used to label the material point of the panel. The inner surface is continuously in contact with an elastic medium that acts as

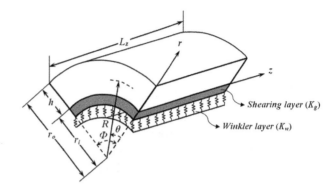

Fig. 2 The sketch of an elastically supported thick bidirectional nanocomposite cylindrical panel resting on a two-parameter elastic foundation and setup of the coordinate system

an elastic foundation represented by the Winkler/Pasternak model with K_w and K_g that are Winkler and shear coefficients of Pasternak foundation, respectively.

One of the well-known power-law distributions which is widely considered by the researchers is three- or four-parameter power-law distribution. The benefit of using such power-law distributions is to illustrate and present useful results arising from symmetric and asymmetric profiles. Consider V_c (volume fraction of the CNTs) in form of $f(z) \times g(r)$, $f(z)$ and $g(r)$ are both the three-parameter power-law distribution. They can be used to illustrate symmetric, asymmetric and classical profiles along the axial and radial directions of the curved panels, respectively. So by considering V_c as $f(z) \times g(r)$, one can present a 2-D six-parameter power-law distribution which is useful to illustrate different types of volume fraction profiles, including classical–classical, symmetric–symmetric and classical–symmetric in both directions.

In order to investigate 3-D dynamic response of thick bidirectional nanocomposite curved panels resting on a two-parameter elastic foundation, it is assumed that the volume fraction of the CNTs follows a 2-D six-parameter power-law distribution:

$$V_c = \left((V_b - V_a)\left(\left(\frac{1}{2} - \frac{r-R}{h} \right) + \alpha_r \left(\frac{1}{2} + \frac{r-R}{h} \right)^{\beta_r} \right)^{\gamma_r} + V_a \right)$$
$$\times \left(1 - \left(\frac{z}{L_z} \right) + \alpha_z \left(\frac{z}{L_z} \right)^{\beta_z} \right)^{\gamma_z}, \qquad (8)$$

where the radial volume fraction index γ_r, and the parameters α_r, β_r and the axial volume fraction index γ_z, and the parameters α_z, β_z govern the material variation profile through the radial and axial directions, respectively. The volume fractions V_a and V_b, which have values that range from 0 to 1, denote the maximum and minimum volume fraction of CNTs. With assumption $V_b = 1$ and $V_a = 0.3$, some material profiles in the radial

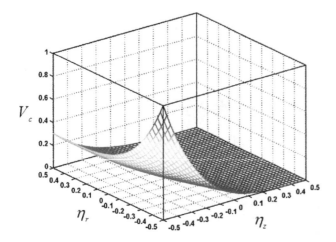

Fig. 3 Variations of the classical volume fraction profile in the radial and axial directions ($\gamma_r = \gamma_z = 4, \alpha_r = \alpha_z = 0$)

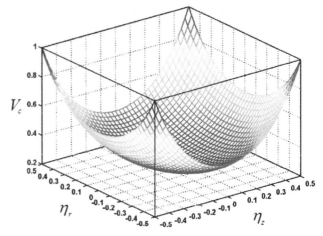

Fig. 5 Variations of the symmetric volume fraction profile along the radial and axial directions of the curved panels ($\gamma_r = \gamma_z = 3, \alpha_r = \alpha_z = 1, \beta_z = \beta_r = 2$)

Table 1 Various volume fraction profiles, different parameters, and volume fraction indices of 2-D power-law distributions

Volume fraction profile	Radial volume fraction index and parameters	Axial volume fraction index and parameters
Classical–classical	$\alpha_r = 0$	$\alpha_z = 0$
Symmetric–symmetric	$\alpha_r = 1, \beta_r = 2$	$\alpha_z = 1, \beta_z = 2$
Classical–symmetric	$\alpha_r = 0$	$\alpha_z = 1, \beta_z = 2$
Classical radially	$\alpha_r = 0$	$\gamma_z = 0$
Symmetric radially	$\alpha_r = 1, \beta_r = 2$	$\gamma_z = 0$

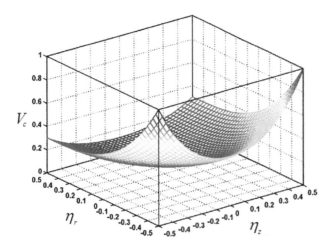

Fig. 4 Variations of the volume fraction profile along the radial and axial directions of the curved panels ($\gamma_r = \gamma_z = 3, \alpha_r = 0, \alpha_z = 1, \beta_z = 2$)

$[\eta_r = (r - R)/h]$ and axial ($\eta_z = z/L_z$) directions are illustrated in Figs. 3, 4 and 5. As can be seen from Fig. 3, the classical volume fraction profiles in the radial and axial directions are presented as special case of the 2-D power-law distribution by setting $\gamma_r = \gamma_z = 4$, and $\alpha_r = \alpha_z = 0$. In Fig. 3, The CNTs volume fraction decreases in the axial direction from 1 at $\eta_z = -0.5$ to 0 at $\eta_z = 0.5$. With another choice of the parameters α_z, β_z, α_r and β_r, it is possible to obtain volume fraction profiles along the radial and axial directions of the panel as shown in Fig. 4. This figure shows a classical profile versus η_r and a symmetric profile versus η_z. As observed, volume fraction on the lower edge ($\eta_z = -0.5$) is the same as that on the upper edge ($\eta_z = 0.5$). Figure 5 illustrates symmetric profiles through the radial and axial directions obtained by setting $\beta_r = \beta_z = 2$, and $\alpha_r = \alpha_z = 1$. In the following, we have

compared several different volume fraction profiles of conventional 1-D and 2-D continuously graded nanocomposite with appropriate choice of the radial and axial parameters of the 2-D six-parameter power-law distribution, as shown in Table 1. It should be noted that the notation classical–symmetric indicates that the 2-D nanocomposite curved panel has classical and symmetric volume fraction profiles in the radial and axial directions, respectively.

The basic formulations

The mechanical constitutive relation that relates the stresses to the strains is as follows:

$$
\begin{bmatrix} \sigma_r \\ \sigma_\theta \\ \sigma_z \\ \tau_{z\theta} \\ \tau_{rz} \\ \tau_{r\theta} \end{bmatrix} = \begin{bmatrix} C_{11} & C_{12} & C_{13} & 0 & 0 & 0 \\ C_{12} & C_{22} & C_{23} & 0 & 0 & 0 \\ C_{13} & C_{23} & C_{33} & 0 & 0 & 0 \\ 0 & 0 & 0 & C_{44} & 0 & 0 \\ 0 & 0 & 0 & 0 & C_{55} & 0 \\ 0 & 0 & 0 & 0 & 0 & C_{66} \end{bmatrix} \begin{bmatrix} \varepsilon_r \\ \varepsilon_\theta \\ \varepsilon_z \\ \gamma_{z\theta} \\ \gamma_{rz} \\ \gamma_{r\theta} \end{bmatrix}.
$$

(9)

In the absence of body forces, the governing equations are as follows:

$$\frac{\partial \sigma_r}{\partial r} + \frac{1}{r}\frac{\partial \tau_{r\theta}}{\partial \theta} + \frac{\partial \tau_{rz}}{\partial z} + \frac{\sigma_r - \sigma_\theta}{r} = \rho\frac{\partial^2 u_r}{\partial t^2},$$

$$\frac{\partial \tau_{r\theta}}{\partial r} + \frac{1}{r}\frac{\partial \sigma_\theta}{\partial \theta} + \frac{\partial \tau_{\theta z}}{\partial z} + \frac{2\tau_{r\theta}}{r} = \rho\frac{\partial^2 u_\theta}{\partial t^2}, \quad (10)$$

$$\frac{\partial \tau_{rz}}{\partial r} + \frac{1}{r}\frac{\partial \tau_{\theta z}}{\partial \theta} + \frac{\partial \sigma_z}{\partial z} + \frac{\tau_{rz}}{r} = \rho\frac{\partial^2 u_z}{\partial t^2}$$

Strain–displacement relations are expressed as:

$$\varepsilon_r = \frac{\partial u_r}{\partial r}, \varepsilon_\theta = \frac{u_r}{r} + \frac{1}{r}\frac{\partial u_\theta}{\partial \theta}, \varepsilon_z = \frac{\partial u_z}{\partial z},$$

$$\gamma_{\theta z} = \frac{\partial u_\theta}{\partial z} + \frac{1}{r}\frac{\partial u_z}{\partial \theta}, \gamma_{rz} = \frac{\partial u_r}{\partial z} + \frac{\partial u_z}{\partial r}, \quad (11)$$

$$\gamma_{r\theta} = \frac{1}{r}\frac{\partial u_r}{\partial \theta} + \frac{\partial u_\theta}{\partial r} - \frac{u_\theta}{r}$$

where u_r, u_θ and u_z are radial, circumferential and axial displacement components, respectively. Upon substitution Eq. (11) into (9) and then into (10), the equations of motion in terms of displacement components with infinitesimal deformations can be written as:

$$c_{11}\frac{\partial^2 u_r}{\partial r^2} + c_{12}\left(-\frac{1}{r^2}\frac{\partial u_\theta}{\partial \theta} + \frac{1}{r}\frac{\partial^2 u_\theta}{\partial r\partial \theta} + \frac{1}{r}\frac{\partial u_r}{\partial r} - \frac{1}{r^2}u_r\right)$$

$$+ c_{13}\frac{\partial^2 u_z}{\partial r\partial z} + \frac{\partial c_{11}}{\partial r}\frac{\partial u_r}{\partial r} + \frac{\partial c_{12}}{\partial r}\left(\frac{u_r}{r} + \frac{1}{r}\frac{\partial u_\theta}{\partial \theta}\right)$$

$$+ \frac{\partial c_{13}}{\partial r}\frac{\partial u_z}{\partial z} + \frac{c_{66}}{r}\left(\frac{\partial^2 u_\theta}{\partial \theta\partial r} + \frac{1}{r}\frac{\partial^2 u_r}{\partial \theta^2} - \frac{1}{r}\frac{\partial u_\theta}{\partial \theta}\right)$$

$$+ c_{55}\left(\frac{\partial^2 u_r}{\partial z^2} + \frac{\partial^2 u_z}{\partial z\partial r}\right) + \frac{1}{r}\left(c_{11}\frac{\partial u_r}{\partial r} + c_{12}(\frac{u_r}{r} + \frac{1}{r}\frac{\partial u_\theta}{\partial \theta})\right.$$

$$\left. + c_{13}\frac{\partial u_z}{\partial z} - c_{12}\frac{\partial u_r}{\partial r} - c_{22}\left(\frac{u_r}{r} + \frac{1}{r}\frac{\partial u_\theta}{\partial \theta}\right) - c_{23}\frac{\partial u_z}{\partial z}\right)$$

$$= \rho\frac{\partial^2 u_r}{\partial t^2} \quad (12)$$

$$c_{66}\left(\frac{-1}{r^2}\frac{\partial u_r}{\partial \theta} + \frac{1}{r}\frac{\partial^2 u_r}{\partial r\partial \theta} + \frac{\partial^2 u_\theta}{\partial r^2} + \frac{u_\theta}{r^2} - \frac{1}{r}\frac{\partial u_\theta}{\partial r}\right)$$

$$+ \frac{\partial c_{66}}{\partial r}\left(\frac{1}{r}\frac{\partial u_r}{\partial \theta} + \frac{\partial u_\theta}{\partial r} - \frac{u_\theta}{r}\right)$$

$$+ \frac{1}{r}\left(c_{12}\frac{\partial^2 u_r}{\partial \theta\partial r} + c_{22}\left(\frac{1}{r}\frac{\partial u_r}{\partial \theta} + \frac{1}{r}\frac{\partial^2 u_\theta}{\partial \theta^2}\right) + c_{23}\frac{\partial^2 u_z}{\partial \theta\partial z}\right)$$

$$+ c_{44}\left(\frac{\partial^2 u_\theta}{\partial z^2} + \frac{1}{r}\frac{\partial^2 u_z}{\partial z\partial \theta}\right)$$

$$+ \frac{2c_{66}}{r}\left(\frac{1}{r}\frac{\partial u_r}{\partial \theta} + \frac{\partial u_\theta}{\partial r} - \frac{u_\theta}{r}\right) = \rho\frac{\partial^2 u_\theta}{\partial t^2} \quad (13)$$

$$c_{55}\left(\frac{\partial^2 u_r}{\partial r\partial z} + \frac{\partial^2 u_z}{\partial r^2}\right) + \frac{\partial c_{55}}{\partial r}\left(\frac{\partial u_r}{\partial z} + \frac{\partial u_z}{\partial r}\right)$$

$$+ \frac{c_{44}}{r}\left(\frac{\partial^2 u_\theta}{\partial \theta\partial z} + \frac{1}{r}\frac{\partial^2 u_z}{\partial \theta^2}\right) + c_{13}\frac{\partial^2 u_r}{\partial z\partial r} + c_{23}\left(\frac{1}{r}\frac{\partial u_r}{\partial z} + \frac{1}{r}\frac{\partial^2 u_\theta}{\partial \theta\partial z}\right)$$

$$+ c_{33}\frac{\partial^2 u_z}{\partial z^2} + \frac{c_{55}}{r}\left(\frac{\partial u_r}{\partial z} + \frac{\partial u_z}{\partial r}\right) = \rho\frac{\partial^2 u_z}{\partial t^2} \quad (14)$$

The boundary conditions at the concave and convex surfaces, $r = r_i$ and r_o, respectively, can be described as follows:

– At $r = r_o$, r_i

$$\tau_{rz} = \tau_{r\theta} = 0, \; \sigma_r$$

$$= \begin{cases} -k_w u_r + k_g\left\{\frac{\partial^2 u_r}{\partial z^2} + \frac{1}{r^2}\frac{\partial^2 u_r}{\partial \theta^2}\right\} & \text{at } r = r_i \\ 0 & \text{at } r = r_o \end{cases} \quad (15)$$

In this investigation, three different types of classical boundary conditions at edges $z = 0$ and L_z of the panel can be stated as follows:

– Simply supported (S):

$$U_r = U_\theta = \sigma_z = 0 \quad (16)$$

– Clamped (C):

$$U_r = U_\theta = U_z = 0 \quad (17)$$

– Free (F):

$$\sigma_z = \sigma_{z\theta} = \sigma_{zr} = 0 \quad (18)$$

For the curved panels with simply supported at one pair of opposite edges, the displacement components can be expanded in terms of trigonometric functions in the direction normal to these edges. In this work, it is assumed that the edges $\theta = 0$ and $\theta = \Phi$ are simply supported. Hence,

$$u_r(r, \theta, z, t) = \sum_{m=1}^{\infty} U_r(r, z)\sin\left(\frac{m\pi}{\Phi}\theta\right)e^{i\omega t},$$

$$u_\theta(r, \theta, z, t) = \sum_{m=1}^{\infty} U_\theta(r, z)\cos\left(\frac{m\pi}{\Phi}\theta\right)e^{i\omega t}, \quad (19)$$

$$u_z(r, \theta, z, t) = \sum_{m=1}^{\infty} U_z(r, z)\sin\left(\frac{m\pi}{\Phi}\theta\right)e^{i\omega t}$$

where m is the circumferential wave number, ω is the natural frequency and $i \, (=\sqrt{-1})$ is the imaginary number. Substituting for displacement components from Eq. (19) into Eqs. (12, 13, 14), one gets

Equation (12):

$$c_{11}\frac{\partial^2 U_r}{\partial r^2} + c_{12}\left(\frac{m\pi}{\Phi r^2}U_\theta - \frac{m\pi}{\Phi r}\frac{\partial U_\theta}{\partial r} + \frac{1}{r}\frac{\partial U_r}{\partial r} - \frac{1}{r^2}U_r\right)$$

$$+ c_{13}\frac{\partial^2 U_z}{\partial r\partial z} + \frac{\partial c_{11}}{\partial r}\frac{\partial U_r}{\partial r} + \frac{\partial c_{12}}{\partial r}\left(\frac{1}{r}U_r - \frac{m\pi}{\Phi r}U_\theta\right) + \frac{\partial c_{13}}{\partial r}\frac{\partial U_z}{\partial z}$$

$$+ \frac{c_{66}}{r}\left(-\frac{m\pi}{\Phi}\frac{\partial U_\theta}{\partial r} - \frac{1}{r}\left(\frac{m\pi}{\Phi}\right)^2 U_r + \frac{m\pi}{\Phi r}U_\theta\right)$$

$$+ c_{55}\left(\frac{\partial^2 U_r}{\partial z^2} + \frac{\partial^2 U_z}{\partial z\partial r}\right) + \frac{1}{r}\left((c_{11}\frac{\partial U_r}{\partial r} + c_{12}\left(\frac{1}{r}U_r - \frac{m\pi}{\Phi r}U_\theta\right)\right)$$

$$+ c_{13}\frac{\partial U_z}{\partial z} - c_{12}\frac{\partial U_r}{\partial r} - c_{22}\left(\frac{1}{r}U_r - \frac{m\pi}{\Phi r}U_\theta\right) - c_{23}\frac{\partial U_z}{\partial z}\right)$$

$$= -\rho\omega^2 U_r \tag{20}$$

Equation (13):

$$c_{66}\left(-\frac{m\pi}{\Phi r^2}U_r + \frac{m\pi}{\Phi r}\frac{\partial U_r}{\partial r} + \frac{\partial^2 U_\theta}{\partial r^2} + \frac{U_\theta}{r^2} - \frac{1}{r}\frac{\partial U_\theta}{\partial r}\right)$$

$$+ \frac{\partial c_{66}}{\partial r}\left(\frac{m\pi}{\Phi r}U_r + \frac{\partial U_\theta}{\partial r} - \frac{U_\theta}{r}\right)$$

$$+ \frac{1}{r}\left(c_{12}\frac{m\pi}{\Phi}\frac{\partial U_r}{\partial r} + c_{22}\left(\frac{m\pi}{\Phi r}U_r - \frac{1}{r}\left(\frac{m\pi}{\Phi}\right)^2 U_\theta\right) + c_{23}\frac{m\pi}{\Phi}\frac{\partial U_z}{\partial z}\right)$$

$$+ c_{44}\left(\frac{\partial^2 U_\theta}{\partial z^2} + \frac{m\pi}{\Phi r}\frac{\partial U_z}{\partial z}\right) + \frac{2c_{66}}{r}\left(\frac{m\pi}{\Phi r}U_r + \frac{\partial U_\theta}{\partial r} - \frac{U_\theta}{r}\right)$$

$$= -\rho\omega^2 U_\theta \tag{21}$$

Equation (14):

$$c_{55}\left(\frac{\partial^2 U_r}{\partial r\partial z} + \frac{\partial^2 U_z}{\partial r^2}\right) + \frac{\partial c_{55}}{\partial r}\left(\frac{\partial U_r}{\partial z} + \frac{\partial U_z}{\partial r}\right)$$

$$+ \frac{c_{44}}{r}\left(-\frac{m\pi}{\Phi}\frac{\partial U_\theta}{\partial z} - \frac{1}{r}\left(\frac{m\pi}{\Phi}\right)^2 U_z\right) + c_{13}\frac{\partial^2 U_r}{\partial z\partial r}$$

$$+ c_{23}\left(\frac{1}{r}\frac{\partial U_r}{\partial z} - \frac{m\pi}{\Phi r}\frac{\partial U_\theta}{\partial z}\right) + c_{33}\frac{\partial^2 U_z}{\partial z^2} + \frac{c_{55}}{r}\left(\frac{\partial U_r}{\partial z} + \frac{\partial U_z}{\partial r}\right)$$

$$= -\rho\omega^2 U_z \tag{22}$$

The geometrical and natural boundary conditions stated in Eq. (15) can also be simplified as follows

Equation (15):

$$\frac{\partial U_r}{\partial z} + \frac{\partial U_z}{\partial r} = \frac{m\pi}{\Phi r}U_r + \frac{\partial U_\theta}{\partial r} - \frac{U_\theta}{r} = 0,$$

$$\begin{cases} c_{11}\frac{\partial U_r}{\partial r} + c_{12}\left(\frac{U_r}{r} - \frac{m\pi}{\Phi r}U_\theta\right) + c_{13}\frac{\partial U_z}{\partial z} \\ \qquad\qquad\qquad\qquad\qquad\qquad\qquad\qquad at\ r = r_i \\ +k_w U_r - k_g\left\{\frac{\partial^2 U_r}{\partial z^2} - \left(\frac{m\pi}{\Phi r}\right)^2 U_r\right\} = 0 \\ c_{11}\frac{\partial U_r}{\partial r} + c_{12}\left(\frac{U_r}{r} - \frac{m\pi}{\Phi r}U_\theta\right) + c_{13}\frac{\partial U_z}{\partial z} = 0 \quad at\ r = r_o \end{cases}$$

$$\tag{23}$$

The boundary conditions stated in Eqs. (16, 17, 18) can also be simplified; however, for the sale of brevity, they are not shown here.

2-D DQM solution of governing equations

It is difficult to solve analytically the equations of motion, if it is not impossible. Hence, one should use an approximate method to find a solution. Here, the differential quadrature method (DQM) is employed. One can compare DQM solution procedure with the other two widely used traditional methods for plate analysis, i.e., Rayleigh–Ritz method and FEM. The main difference between the DQM and the other methods is how the governing equations are discretized. In DQM, the governing equations and boundary conditions are directly discretized, and thus elements of stiffness and mass matrices are evaluated directly. But in Rayleigh–Ritz and FEMs, the weak form of the governing equations should be developed and the boundary conditions are satisfied in the weak form. Generally by doing so larger number of integrals with increasing amount of differentiation should be done to arrive at the element matrices. In addition, the number of degrees of freedom will be increased for an acceptable accuracy.

The basic idea of the DQM is the derivative of a function, with respect to a space variable at a given sampling point, is approximated as a weighted linear sum of the sampling points in the domain of that variable. In order to illustrate the DQ approximation, consider a function $f(\xi, \eta)$ defined on a rectangular domain $0 \le \xi \le a$ and $0 \le \eta \le b$. Let in the given domain, the function values be known or desired on a grid of sampling points. According to DQM method, the rth derivative of the function $f(\xi, \eta)$ can be approximated as:

$$\frac{\partial^r f(\xi, \eta)}{\partial \xi^r}\Big|(\xi, \eta) = (\xi_i, \eta_j)$$

$$= \sum_{m=1}^{N_\xi} A_{im}^{\xi(r)} f_{mj} \quad for\ i = 1, 2, \dots, N_\xi \tag{24}$$

$$and\ r = 1, 2, \dots, N_\xi - 1$$

where N_ξ represents the total number of nodes along the ξ-direction. From this equation one can deduce that the important components of DQM approximations are the weighting coefficients $(A_{ij}^{\xi(r)})$ and the choice of sampling points. In order to determine the weighting coefficients, a set of test functions should be used in Eq. (24). The weighting coefficients for the first-order derivatives in ξ-direction are thus determined as (Bellman and Casti 1971):

$$A_{ij}^{\xi} = \begin{cases} \dfrac{1}{a}\dfrac{M(\xi_i)}{(\xi_i - \xi_j)M(\xi_j)} & \text{for } i \neq j \\ -\sum\limits_{\substack{j=1 \\ i \neq j}}^{N_\xi} A_{ij}^{\xi} & \text{for } i=j \end{cases} ; \quad i,j = 1,2,\ldots,N_\xi, \tag{25}$$

where

$$M(\xi_i) = \prod_{j=1, i \neq j}^{N_\xi} (\xi_i - \xi_j) \tag{26}$$

$j = 1, 2, \ldots, N_\eta$, where N_ξ and N_η are the total number of nodes along the ξ- and η-directions, respectively. At this stage, the DQ method can be applied to discretize the equations of motion (20, 21, 22). As a result, at each domain grid point (r_i, z_j) with $i = 2, \ldots, N_r - 1$ and $j = 2, \ldots, N_z - 1$, the discretized equations take the following forms.

Equation (20):

$$(c_{11})_{ij} \sum_{n=1}^{N_r} B_{in}^r U_{rnj} + (c_{12})_{ij} \left(\frac{m\pi}{\Phi r_i^2} U_{\theta ij} - \frac{m\pi}{\Phi r_i} \sum_{n=1}^{N_r} A_{in}^r U_{\theta nj} + \frac{1}{r_i} \sum_{n=1}^{N_r} A_{in}^r U_{rnj} - \frac{U_{rij}}{r_i^2} \right)$$

$$+ (c_{13})_{ij} \sum_{n=1}^{N_r} \sum_{v=1}^{N_z} A_{jv}^z A_{in}^r U_{znv} + \left(\frac{\partial c_{11}}{\partial r} \right) \sum_{ij}^{N_r} A_{in}^r U_{rnj} + \left(\frac{\partial c_{12}}{\partial r} \right)_{ij} \left(\frac{1}{r_i} U_{rij} - \frac{m\pi}{\Phi r_i} U_{\theta ij} \right)$$

$$+ \left(\frac{\partial c_{13}}{\partial r} \right) \sum_{ij}^{N_z} A_{jn}^z U_{zin} + \frac{(c_{66})_{ij}}{r_i} \left(-\frac{m\pi}{\Phi} \sum_{n=1}^{N_r} A_{in}^r U_{\theta nj} - \frac{1}{r_i} \left(\frac{m\pi}{\Phi} \right)^2 U_{rij} + \frac{m\pi}{\Phi r_i} U_{\theta ij} \right)$$

$$+ (c_{55})_{ij} \left(\sum_{n=1}^{N_z} B_{jn}^z U_{rin} + \sum_{n=1}^{N_r} \sum_{v=1}^{N_z} A_{jv}^z A_{in}^r U_{znv} \right)$$

$$+ \frac{1}{r_i} \left((c_{11})_{ij} \sum_{n=1}^{N_r} A_{in}^r U_{rnj} + (c_{12})_{ij} \left(\frac{U_{rij}}{r_i} - \frac{m\pi}{\Phi r_i} U_{\theta ij} \right) + (c_{13})_{ij} \sum_{n=1}^{N_z} A_{jn}^z U_{zin} - (c_{12})_{ij} \sum_{n=1}^{N_r} A_{in}^r U_{rnj} \right.$$

$$\left. - (c_{22})_{ij} \left(\frac{U_{rij}}{r_i} - \frac{m\pi}{\Phi r_i} U_{\theta ij} \right) - (c_{23})_{ij} \sum_{n=1}^{N_z} A_{jn}^z U_{zin} \right)$$

$$= -\rho_{ij} \omega^2 U_{rij} \tag{29}$$

The weighting coefficients of the second-order derivative can be obtained in the matrix form (Bellman and Casti 1971):

$$\left[B_{ij}^\xi \right] = \left[A_{ij}^\xi \right] \left[A_{ij}^\xi \right] = \left[A_{ij}^\xi \right]^2 \tag{27}$$

In a similar manner, the weighting coefficients for the η-direction can be obtained.

The natural and simplest choice of the grid points is equally spaced points in the direction of the coordinate axes of computational domain. It was demonstrated that non-uniform grid points gives a better result with the same number of equally spaced grid points (Bellman and Casti 1971). It is shown (Shu and Wang 1999) that one of the best options for obtaining grid points is Chebyshev–Gauss–Lobatto quadrature points:

$$\frac{\xi_i}{a} = \frac{1}{2} \left\{ 1 - \cos \left[\frac{(i-1)\pi}{(N_\xi - 1)} \right] \right\},$$

$$\frac{\eta_j}{b} = \frac{1}{2} \left\{ 1 - \cos \left[\frac{(j-1)\pi}{(N_\eta - 1)} \right] \right\} \quad \text{for } i = 1,2,\ldots,N_\xi;$$

$$\tag{28.1, 2}$$

Equation (21):

$$(c_{66})_{ij} \left(-\frac{1}{r_i^2} \frac{m\pi}{\Phi} U_{rij} + \frac{m\pi}{\Phi r_i} \sum_{n=1}^{N_r} A_{in}^r U_{rnj} \right.$$

$$\left. + \sum_{n=1}^{N_r} B_{in}^r U_{\theta nj} + \frac{U_{\theta ij}}{r_i^2} - \frac{1}{r_i} \sum_{n=1}^{N_r} A_{in}^r U_{\theta nj} \right)$$

$$+ \left(\frac{\partial c_{66}}{\partial r} \right)_{ij} \left(\frac{m\pi}{\Phi r_i} U_{rij} + \sum_{n=1}^{N_r} A_{in}^r U_{\theta nj} - \frac{U_{\theta ij}}{r_i} \right)$$

$$+ \frac{1}{r_i} \left((c_{12})_{ij} \frac{m\pi}{\Phi} \sum_{n=1}^{N_r} A_{in}^r U_{rnj} + (c_{22})_{ij} \left(\frac{m\pi}{\Phi r_i} U_{rij} - \frac{1}{r_i} \left(\frac{m\pi}{\Phi} \right)^2 U_{\theta ij} \right) \right.$$

$$\left. + (c_{23})_{ij} \frac{m\pi}{\Phi} \sum_{n=1}^{N_z} A_{jn}^z U_{zin} \right)$$

$$+ (c_{44})_{ij} \left(\sum_{n=1}^{N_z} B_{jn}^z U_{\theta in} + \frac{m\pi}{\Phi r_i} \sum_{n=1}^{N_z} A_{jn}^z U_{zin} \right)$$

$$+ \frac{2(c_{66})_{ij}}{r_i} \left(\frac{m\pi}{\Phi r_i} U_{rij} + \sum_{n=1}^{N_r} A_{in}^r U_{\theta nj} - \frac{U_{\theta ij}}{r_i} \right)$$

$$= -\rho_{ij} \omega^2 U_{\theta ij} \tag{30}$$

Equation (22):

$$
(c_{55})_{ij} \left(\sum_{n=1}^{N_r} \sum_{v=1}^{N_z} A_{jv}^z A_{in}^r U_{rnv} + \sum_{n=1}^{N_r} B_{in}^r U_{znj} \right) + \left(\frac{\partial c_{55}}{\partial r} \right)_{ij}
$$

$$
\times \left(\sum_{n=1}^{N_z} A_{jn}^z U_{rin} + \sum_{n=1}^{N_r} A_{in}^r U_{znj} \right)
$$

$$
+ \frac{(c_{44})_{ij}}{r_i} \left(-\frac{m\pi}{\Phi} \sum_{n=1}^{N_z} A_{jn}^z U_{\theta in} - \frac{1}{r_i} \left(\frac{m\pi}{\Phi} \right)^2 U_{zij} \right)
$$

$$
+ (c_{13})_{ij} \sum_{n=1}^{N_r} \sum_{v=1}^{N_z} A_{jv}^z A_{in}^r U_{rnv}
$$

$$
+ (c_{23})_{ij} \left(\frac{1}{r_i} \sum_{n=1}^{N_z} A_{jn}^z U_{rin} - \frac{m\pi}{\Phi r_i} \sum_{n=1}^{N_z} A_{jn}^z U_{\theta in} \right)
$$

$$
+ (c_{33})_{ij} \sum_{n=1}^{N_z} B_{jn}^z U_{zin} + \frac{(c_{55})_{ij}}{r_i} \left(\sum_{n=1}^{N_z} A_{jn}^z U_{rin} + \sum_{n=1}^{N_r} A_{in}^r U_{znj} \right)
$$

$$
= -\rho_{ij} \omega^2 U_{zij} \tag{31}
$$

where A_{ij}^r, A_{ij}^z and B_{ij}^r, B_{ij}^z are the first- and second-order DQ weighting coefficients in the r- and z-directions, respectively. The DQ method can also be applied to discretize the boundary conditions at $r = r_i$ and r_o as follows. Equation (23):

$$
\sum_{n=1}^{N_z} A_{jn}^z U_{rin} + \sum_{n=1}^{N_r} A_{in}^r U_{znj} = 0,
$$

$$
\frac{m\pi}{\Phi r_i} U_{rij} + \sum_{n=1}^{N_r} A_{in}^r U_{\theta nj} - \frac{U_{\theta ij}}{r_i} = 0,
$$

$$
(c_{11})_{ij} \sum_{n=1}^{N_r} A_{in}^r U_{rnj} + (c_{12})_{ij} \left(\frac{U_{rij}}{r_i} - \frac{m\pi}{\Phi r_i} U_{\theta ij} \right)
$$

$$
+ (c_{13})_{ij} \sum_{n=1}^{N_z} A_{jn}^z U_{zin}
$$

$$
\left\{ k_w U_{rij} - k_g \left(\sum_{n=1}^{N_z} B_{jn}^z U_{rin} - \left(\frac{m\pi}{\Phi r_i} \right)^2 U_{rij} \right) \right\} \delta_{1i} = 0 \tag{32}
$$

where $i = 1$ at $r = r_i$ and $i = N_r$ at $r = r_o$, and $j = 1, 2, \ldots, N_z$; also δ_{ij} is the Kronecker delta. The boundary conditions at $z = 0$ and L_z stated in Eqs. (16, 17, 18), become Eq. (16):

• Simply supported (S):

$$
U_{rij} = U_{\theta ij} = 0,
$$

$$
(c_{13})_{ij} \sum_{n=1}^{N_r} A_{in}^r U_{rnj} + (c_{23})_{ij} \left(\frac{U_{rij}}{r_i} - \frac{m\pi}{\Phi r_i} U_{\theta ij} \right)
$$

$$
+ (c_{33})_{ij} \sum_{n=1}^{N_z} A_{jn}^z U_{zin} = 0 \tag{33}
$$

Equation (17):

• Clamped (C):

$$
U_{rij} = U_{\theta ij} = U_{zij} 0 \tag{34}
$$

Equation (18):

• Free (F):

$$
(c_{13})_{ij} \sum_{n=1}^{N_r} A_{in}^r U_{rnj} + (c_{23})_{ij} \left(\frac{U_{rij}}{r_i} - \frac{m\pi}{\Phi r_i} U_{\theta ij} \right)
$$

$$
+ (c_{33})_{ij} \sum_{n=1}^{N_z} A_{jn}^z U_{zin} = 0,
$$

$$
\sum_{n=1}^{N_z} A_{jn}^z U_{\theta in} + \frac{m\pi}{\Phi r_i} U_{zij} = 0, \tag{35}
$$

$$
\sum_{n=1}^{N_z} A_{jn}^z U_{rin} + \sum_{n=1}^{N_r} A_{in}^r U_{znj} = 0
$$

In the above equations $i = 2, \ldots, N_r - 1$; also $j = 1$ at $z = 0$ and $j = N_z$ at $z = L_z$.

In order to carry out the eigenvalue analysis, the domain and boundary nodal displacements should be separated. In vector forms, they are denoted as $\{d\}$ and $\{b\}$, respectively. Based on this definition, the discretized form of the equations of motion and the related boundary conditions can be represented in the matrix form as:Equations of motion, Eqs. (29, 30, 31):

$$
[[K_{db}][K_{dd}]] \left\{ \begin{array}{c} \{b\} \\ \{d\} \end{array} \right\} - \omega^2 [M]\{d\} = \{0\} \tag{36}
$$

Boundary conditions, Eq. (32) and Eqs. (33, 34, 35):

$$
[K_{bd}]\{d\} + [K_{bb}]\{b\} = \{0\} \tag{37}
$$

Eliminating the boundary degrees of freedom in Eq. (36) using Eq. (37), this equation becomes

$$
[K] - \omega^2 [M]\{d\} = \{0\}, \tag{38}
$$

where $[K] = [K_{dd}] - [K_{db}][K_{bb}]^{-1}[K_{bd}]$. The above eigenvalue system of equations can be solved to find the natural frequencies and mode shapes of the curved panel.

Numerical results and discussion

Convergence and comparison studies

Due to lack of appropriate results for free vibration of CGCNTR cylindrical panels reinforced by oriented CNTs for direct comparison, validation of the presented formulation is conducted in two ways. Firstly, the results are compared with those of FGM composite cylindrical panels and then, the results of the presented formulations are given in the form of convergence studies with respect to N_r and N_z, the number of discrete points distributed along the radial and axial directions, respectively. To validate the proposed approach its convergence and accuracy are demonstrated via different examples. The obtained natural frequencies based on the three-dimensional elasticity formulation are compared with those of the power series expansion method for both FGM curved panels with and without elastic foundations (Matsunaga 2008; Pradyumna and Bandyopadhyay 2008; Farid et al. 2010). In these studies the material properties of functionally graded materials are assumed as follows:

- Metal (Aluminum, Al):

  $E_m = 70 * 10^9$ Pa, $\rho_m = 2702$ Kg/m^3, $\upsilon_m = 0.3$

- Ceramic (Alumina, Al$_2$O$_3$):

  $E_c = 380 * 10^9$ Pa, $\rho_c = 3800$ Kg/m^3, $\upsilon_c = 0.3$

Subscripts M and C refer to the metal and ceramic constituents which denote the material properties of the outer and inner surfaces of the panel, respectively. To validate the analysis, results for FGM cylindrical shells are compared with similar ones in the literature, as shown in Table 2. The comparison shows that the present results agreed well with those in the literatures. Besides the fast

Table 2 Comparison of the normalized natural frequency of an FGM composite curved panel with four edges simply supported $(\Omega_{11} = \omega_{11} R\Phi \sqrt{\rho_m h/D}, D = E_m h^3 / 12(1 - \upsilon_m^2))$

P (volume fraction index)		R/L_z				
		0.5	1	5	10	50
0	$N_r = N_z = 5$	69.9774	52.1052	42.7202	42.3717	42.2595
	$N_r = N_z = 7$	69.9722	52.1052	42.7158	42.3718	42.2550
	$N_r = N_z = 9$	69.9698	52.1003	42.7159	42.3700	42.2553
	$N_r = N_z = 11$	69.9700	52.1003	42.7160	42.3677	42.2552
	$N_r = N_z = 13$	69.9700	52.1003	42.7160	42.3677	42.2553
	Pradyumna and Bandyopadhyay (2008)	68.8645	51.5216	42.2543	41.908	41.7963
0.2	$N_r = N_z = 5$	65.1470	47.9393	39.1282	38.8010	38.7020
	$N_r = N_z = 7$	65.4449	48.0456	39.1008	38.7366	38.6834
	$N_r = N_z = 9$	65.4526	48.1340	39.0836	38.7568	38.6581
	$N_r = N_z = 11$	65.4304	48.1340	39.0835	38.7568	38.6580
	$N_r = N_z = 13$	65.4304	48.1340	39.0835	38.7568	38.6581
	Pradyumna and Bandyopadhyay (2008)	64.4001	47.5968	40.1621	39.8472	39.7465
0.5	$N_r = N_z = 5$	60.1196	43.5539	36.1264	35.8202	34.7341
	$N_r = N_z = 7$	60.2769	43.7128	36.1401	35.7964	35.0677
	$N_r = N_z = 9$	60.3574	43.7689	36.0944	35.7890	35.7032
	$N_r = N_z = 11$	60.3574	43.7688	36.0943	35.7891	35.7032
	$N_r = N_z = 13$	60.3574	43.7689	36.0944	35.7891	35.7032
	Pradyumna and Bandyopadhyay (2008)	59.4396	43.3019	37.287	36.9995	36.9088
1	$N_r = N_z = 5$	54.1034	38.5180	31.9860	30.7065	30.6336
	$N_r = N_z = 7$	54.6039	39.1477	32.1140	31.6982	31.5397
	$N_r = N_z = 9$	54.7141	39.1620	32.0401	31.7608	31.6877
	$N_r = N_z = 11$	54.7141	39.1621	32.0401	31.7608	31.6878
	$N_r = N_z = 13$	54.7141	39.1621	32.0401	31.7608	31.6877
	Pradyumna and Bandyopadhyay (2008)	53.9296	38.7715	33.2268	32.9585	32.875
2	$N_r = N_z = 5$	46.9016	34.7702	27.6657	27.4295	27.3725
	$N_r = N_z = 7$	47.9865	34.6980	27.5733	27.3389	27.2669
	$N_r = N_z = 9$	48.5250	34.6852	27.5614	27.3238	27.2663
	$N_r = N_z = 11$	48.5250	34.6851	27.5614	27.3239	27.2663
	$N_r = N_z = 13$	48.5250	34.6851	27.5614	27.3239	27.2662
	Pradyumna and Bandyopadhyay (2008)	47.8259	34.3338	27.4449	27.1789	27.0961

rate of convergence of the method being quite evident, it is found that only 13 grid points ($N_r = N_z = 13$) along the radial and axial directions can yield accurate results. Further validation of the present results for isotropic FGM cylindrical panel is shown in Table 3. In this table, comparison is made for different L_z/R and L_z/h ratios, and it is observed there is good agreement between the results.

As another example, the convergence and accuracy of the method are investigated by evaluating the first three natural frequency parameters of the FG curved panel resting on Pasternak foundations. The non-dimensional forms of the elastic foundation coefficients are defined as $K_w = k_w R/G_c$ and $K_g = k_g/(G_c R)$ in which G_c is the shear modulus of elasticity of the ceramic layer. The results are prepared for different thickness-to-mean radius ratios and different values of the DQ grid points along the radial and axial

directions, respectively, are shown in Table 4. Also, one can see that excellent agreement exists between the results.

Parametric studies

After demonstrating the convergence and accuracy of the present method, parametric studies for 3-D vibration analysis of thick curved panels resting on a two-parameter elastic foundation reinforced by randomly oriented straight single-walled carbon nanotubes for various CNTs volume fraction distribution, length-to-mean radius ratio, elastic coefficients of foundation and different combinations of free, simply supported and clamped boundary conditions along the axial direction of the curved panel, are computed. The boundary conditions of the panel are specified by the letter symbols, for example, S–C–S–F denotes a curved

Table 3 Comparison of the normalized natural frequency of an FGM composite curved panel for various L_z/R and L_z/h ratios

			P (volume fraction index)				
			0	0.5	1	4	10
$L_z/h = 2$		$L_z/R = 0.5$					
	Matsunaga (2008)		0.9334	0.8213	0.7483	0.6011	0.5461
	Farid et al. (2010)		0.9187	0.8013	0.7263	0.5267	0.5245
	$N_r = N_z = 5$		0.9342	0.8001	0.7149	0.5878	0.5133
	$N_r = N_z = 7$		0.9249	0.8011	0.7250	0.5783	0.5298
	$N_r = N_z = 9$		0.9250	0.8018	0.7253	0.5790	0.5301
	$N_r = N_z = 11$		0.9249	0.8017	0.7253	0.5789	0.5300
	$N_r = N_z = 13$		0.9250	0.8018	0.7252	0.5790	0.5301
	Matsunaga (2008)	$L_z/R = 1$	0.9163	0.8105	0.7411	0.5967	0.5392
	Farid et al. (2010)		0.8675	0.7578	0.6875	0.5475	0.4941
	$N_r = N_z = 5$		0.8942	0.7531	0.6746	0.5741	0.4913
	$N_r = N_z = 7$		0.8851	0.7671	0.6912	0.5599	0.5074
	$N_r = N_z = 9$		0.8857	0.7666	0.6935	0.5531	0.5065
	$N_r = N_z = 11$		0.8857	0.7667	0.6934	0.5531	0.5063
	$N_r = N_z = 13$		0.8856	0.7667	0.6935	0.5532	0.5064
$L_z/h = 5$		$L_z/R = 0.5$					
	Matsunaga (2008)		0.2153	0.1855	0.1678	0.1413	0.1328
	Farid et al. (2010)		0.2113	0.1814	0.1639	0.1367	0.1271
	$N_r = N_z = 5$		0.2230	0.1997	0.1542	0.1374	0.1373
	$N_r = N_z = 7$		0.2176	0.1823	0.1624	0.1362	0.1233
	$N_r = N_z = 9$		0.2130	0.1817	0.1639	0.1374	0.1296
	$N_r = N_z = 11$		0.2128	0.1816	0.1640	0.1377	0.1296
	$N_r = N_z = 13$		0.2129	0.1817	0.1640	0.1374	0.1295
	Matsunaga (2008)	$L_z/R = 1$	0.2239	0.1945	0.1769	0.1483	0.1385
	Farid et al. (2010)		0.2164	0.1879	0.1676	0.1394	0.1286
	$N_r = N_z = 5$		0.2066	0.1765	0.1567	0.1476	0.1409
	$N_r = N_z = 7$		0.2133	0.1843	0.1688	0.1377	0.1288
	$N_r = N_z = 9$		0.2154	0.1848	0.1671	0.1392	0.1301
	$N_r = N_z = 11$		0.2155	0.1847	0.1675	0.1392	0.1299
	$N_r = N_z = 13$		0.2155	0.1847	0.1671	0.1392	0.1302

Table 4 Comparison of the first three non-dimensional natural frequency parameters of panel on an elastic foundation ($\varpi_{mn} = \omega_{mn}h\sqrt{\rho_C/E_C}, P = 1, \Phi = 60°, N_r = N_z = 13$)

L_z/R	h/R	K_w, K_g		ϖ_{11}	ϖ_{22}	ϖ_{33}
1	0.1	1, 0.1	Present	0.2201	0.4411	0.6462
			Farid et al. (2010)	0.2200	0.4403	0.6427
		100, 10	Present	0.2241	0.4475	0.6679
			Farid et al. (2010)	0.2243	0.4475	0.6681
	0.5	1, 0.1	Present	0.8043	1.8601	2.9796
			Farid et al. (2010)	0.8041	1.8599	2.9796
		100, 10	Present	0.9500	1.8964	2.9956
			Farid et al. (2010)	0.9503	1.8963	2.9956
2	0.1	1, 0.1	Present	0.1715	0.3430	0.5121
			Farid et al. (2010)	0.1712	0.3434	0.5122
		100, 10	Present	0.174	0.3477	0.5202
			Farid et al. (2010)	0.174	0.3475	0.5200
	0.5	1, 0.1	Present	0.5769	1.3408	2.1825
			Farid et al. (2010)	0.5772	1.3409	2.1827
		100, 10	Present	0.7664	1.4034	2.2027
			Farid et al. (2010)	0.7664	1.4037	2.2023

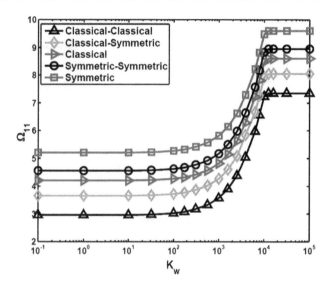

Fig. 6 Variations of fundamental frequency parameters of a bidirectional S–C–S–C nanocomposite curved panels resting on a two-parameter elastic foundation with Winkler elastic coefficient for different volume fraction profiles ($K_g = 100$, $R/h = L_z/R = 3.5$, $\gamma_r = 2$, $\Phi = 135°$)

panel with edges $\theta = 0$ and Φ simply supported (S), edge $z = 0$ clamped (C) and edge $z = L_z$ free (F).

The non-dimensional natural frequency, Winkler and shearing layer elastic coefficients are as follows:

$$\Omega_{mn} = \omega_{mn}10h\sqrt{\rho_m/E_m},$$
$$K_w = \frac{k_wR}{G_m}, K_g = \frac{k_g}{G_mR}, \tag{39}$$

where ρ_m, E_m and G_m represent the mass density, Young's modulus and shear modulus of the matrix, respectively.

The effect of the Winkler elastic coefficient on the fundamental frequency parameters for different boundary conditions is shown in Figs. 6, 7 and 8. It is observed that the fundamental frequency parameters converge with increasing Winkler elastic coefficient of the foundation. According to these figures, the lowest frequency parameter is obtained by using classical–classical volume fraction profile. On the contrary, the 1-D FG panel with symmetric volume fraction profile has the maximum value of the frequency parameter. Therefore, a graded CNTs volume fraction in two directions has higher capabilities to reduce the frequency parameter than conventional 1-D nanocomposite. It is also observed from Figs. 6, 7 and 8, for the large values of Winkler elastic coefficient, the shearing layer elastic coefficient has less effect and the results become independent of it, in other words the non-dimensional natural frequencies converge with increasing Winkler foundation stiffness.

The influence of shearing layer elastic coefficient on the non-dimensional natural frequency for S–C–S–C, S–C–S–

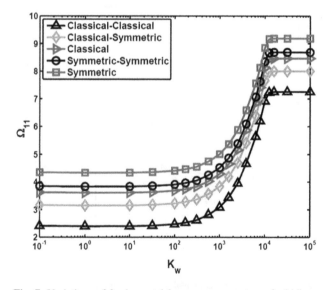

Fig. 7 Variations of fundamental frequency parameters of a bidirectional S–C–S–S nanocomposite curved panels resting on a two-parameter elastic foundation with Winkler elastic coefficient for different volume fraction profiles ($K_g = 100$, $R/h = L_z/R = 3.5$, $\gamma_r = 2$, $\Phi = 135°$)

S and S–F–S–F bidirectional nanocomposite curved panel resting on a two-parameter elastic foundation, is shown in Figs. 9, 10 and 11. It is observed that the variation of Winkler elastic coefficient has little effect on the non-dimensional natural frequency at different values of shearing layer elastic coefficient. It is clear that with increasing the shearing layer elastic coefficient of the foundation, the frequency parameters increase to some limit values and for

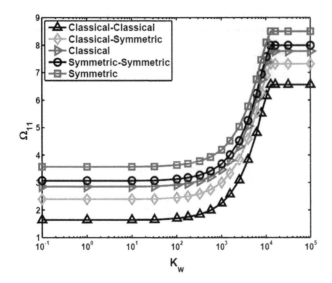

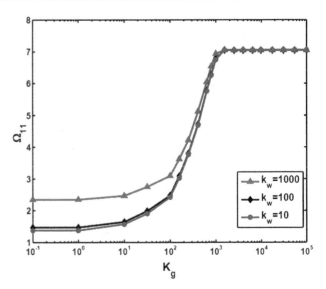

Fig. 8 Variations of fundamental frequency parameters of a bidirectional S–F–S–F nanocomposite curved panels resting on a two-parameter elastic foundation with Winkler elastic coefficient for different volume fraction profiles ($K_g = 100$, $R/h = L_z/R = 3.5$, $\gamma_r = 2$, $\Phi = 135°$)

Fig. 10 Variations of fundamental frequency parameters of a bidirectional S–C–S–S nanocomposite curved panels resting on a two-parameter elastic foundation with the shearing layer elastic coefficient ($R/h = L_z/R = 3.5$, $\gamma_r = \gamma_z = 2$, $\alpha_r = \alpha_z = 0$, $\Phi = 135°$)

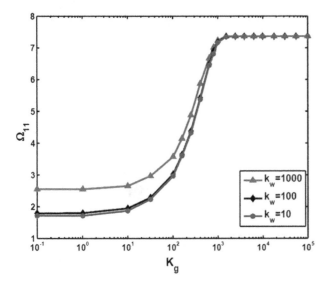

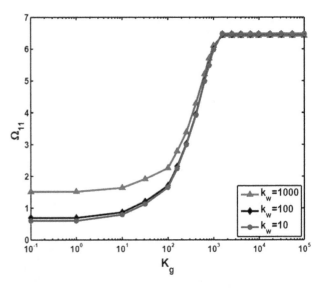

Fig. 9 Variations of fundamental frequency parameters of a bidirectional S–C–S–C nanocomposite curved panels resting on a two-parameter elastic foundation with the shearing layer elastic coefficient ($R/h = L_z/R = 3.5$, $\gamma_r = \gamma_z = 2$, $\alpha_r = \alpha_z = 0$, $\Phi = 135°$)

Fig. 11 Variations of fundamental frequency parameters of a bidirectional S–F–S–F nanocomposite curved panels resting on a two-parameter elastic foundation with the shearing layer elastic coefficient ($R/h = L_z/R = 3.5$, $\gamma_r = \gamma_z = 2$, $\alpha_r = \alpha_z = 0$, $\Phi = 135°$)

the large values of shearing layer elastic coefficient, the frequency parameters become independent of it.

The variations of fundamental frequency parameters of bidirectional nanocomposite curved panels resting on an elastic foundation with length-to-mean radius ratio (L_z/R) for different types of volume fraction profiles are depicted in Figs. 12, 13 and 14. It can also be inferred from these figures that the frequency is greatly influenced in that fundamental frequency parameter decreases steadily as

length-to-mean radius ratio (L_z/R) becomes larger and remains almost unaltered for the large values of length-to-mean radius ratio. As can be seen from this figure, for the all length-to-mean radius ratio (L_z/R), classical–classical volume fraction profile has the lowest frequencies followed by classical–symmetric, classical, symmetric–symmetric and symmetric profiles.

The variations of fundamental frequency parameters of bidirectional nanocomposite curved panels with length-to-

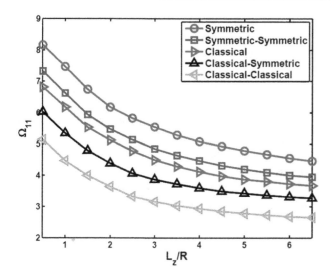

Fig. 12 Variations of fundamental frequency parameters of two-dimensional continuously graded S–C–S–C nanocomposite curved panels resting on an elastic foundation with L_z/R ratio for different volume fraction profiles ($K_w = K_g = 100$, $R/h = 3.5$, $\gamma_z = 2$, $\Phi = 135°$)

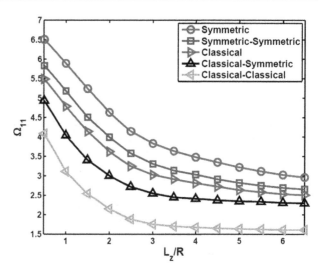

Fig. 14 Variations of fundamental frequency parameters of two-dimensional continuously graded S–F–S–F nanocomposite curved panels resting on an elastic foundation with L_z/R ratio for different volume fraction profiles ($K_w = K_g = 100$, $R/h = 3.5$, $\gamma_z = 2$, $\Phi = 135°$)

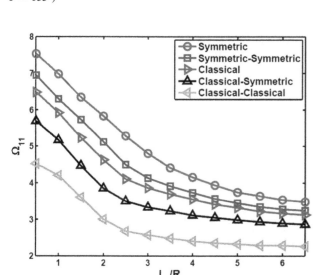

Fig. 13 Variations of fundamental frequency parameters of two-dimensional continuously graded S–C–S–S nanocomposite curved panels resting on an elastic foundation with L_z/R ratio for different volume fraction profiles ($K_w = K_g = 100$, $R/h = 3.5$, $\gamma_z = 2$, $\Phi = 135°$)

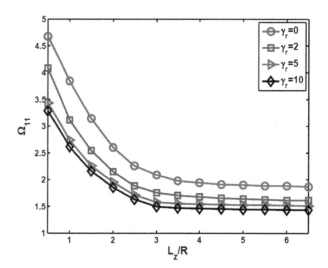

Fig. 15 Variations of fundamental frequency parameters of bidirectional nanocomposite curved panels with length-to-mean radius ratio (L_z/R), and the volume fraction index through the radial direction of the panels for S–F–S–F boundary condition ($K_w = K_g = 100$, $R/h = 3.5$, $\gamma_z = 2$, $\alpha_r = \alpha_z = 0$, $\Phi = 135°$)

mean radius ratio (L_z/R), and the volume fraction index through the radial direction of the panels for $S–F–S–F$ boundary conditions are shown in Fig. 15, by considering ($\alpha_r = \alpha_z = 0, \gamma_z = 2, K_w = K_g = 100$) for classical–classical 2-D nanocomposite curved panels. Confirming the effect of length-to-mean radius ratio on the natural frequency already shown in the Figs. 12, 13 and 14, it is found that the frequency parameter decreases by increasing the radial volume fraction index (γ_r). This behavior is also

observed for other boundary conditions, not shown here for brevity.

Conclusion remarks

In this research work, free vibration of thick bidirectional nanocomposite curved panels resting on a two-parameter elastic is investigated based on three-dimensional theory of elasticity. The elastic foundation is considered as a Pasternak model with adding a shear layer to the Winkler model. Three complicated equations of motion for the curved panel under consideration are semi-analytically solved by using 2-D differential quadrature method. Using the 2-D differential quadrature method along the radial and axial directions, allows one to deal with curved panel with arbitrary thickness distribution of material properties and also to implement the effects of the elastic foundations as a boundary condition on the lower surface of the curved panel efficiently and in an exact manner. The volume fractions of randomly oriented straight single-walled carbon nanotubes (SWCNTs) are assumed to be graded not only in the radial direction, but also in axial direction of the curved panel. The direct application of CNTs properties in micromechanics models for predicting material properties of the nanotube/polymer composite is inappropriate without taking into account the effects associated with the significant size difference between a nanotube and a typical carbon fiber. In other words, continuum micromechanics equations cannot capture the scale difference between the nano and micro-levels. In order to overcome this limitation, a virtual equivalent fiber consisting of nanotube and its inter-phase which is perfectly bonded to surrounding resin is applied. In this research work, an equivalent continuum model based on the Eshelby–Mori–Tanaka approach is employed to estimate the effective constitutive law of the elastic isotropic medium (matrix) with oriented straight CNTs. The effects of elastic foundation stiffness parameters, various geometrical parameters on the vibration characteristics of CGCNTR curved panel, are investigated, and also, different types of volume fraction profiles along the radial and axial directions of the panels and elastic coefficients of foundation of bidirectional curved panels resting on a two-parameter elastic foundation are studied. Moreover, vibration behavior of 2-D continuously graded nanocomposite panels are compared with conventional one-dimensional nanocomposite panels. From this study, some conclusions can be made:

- It is observed, for the large values of Winkler elastic coefficient, the shearing layer elastic coefficient has less effect and the results become independent of it, in other words the non-dimensional natural frequencies converge with increasing Winkler foundation stiffness.
- The results show that the variation of Winkler elastic coefficient has little effect on the non-dimensional natural frequency at different values of shearing layer elastic coefficient. It is clear that with increasing the

shearing layer elastic coefficient of the foundation, the frequency parameters increase to some limit values and for the large values of shearing layer elastic coefficient; the frequency parameters become independent of it.

- The frequency parameter decreases rapidly with the increase of the length-to-mean radius ratio and then remains almost unaltered for the long cylindrical panel ($L_z/R > 5$).
- The interesting results show that the lowest magnitude frequency parameter is obtained by using a classical–classical volume fraction profile. It can be concluded that a graded CNTs volume fraction in two directions has higher capabilities to reduce the natural frequency than a conventional 1-D nanocomposite.
- It is found that the frequency parameter decreases by increasing the radial volume fraction index (γ_r).
- For the all length-to-mean radius ratio (L_z/R), classical–classical volume fraction profile has the lowest frequencies followed by classical–symmetric, classical, symmetric–symmetric and symmetric profiles.

Based on the achieved results, using 2-D six-parameter power-law distribution leads to a more flexible design so that maximum or minimum value of natural frequency can be obtained in a required manner.

Appendix

The Hill's elastic moduli are found as (Shi et al. 2004):

$$\varepsilon_r = k = \frac{E_m\{E_m f_m + 2k_r(1+v_m)[1+f_r(1-2v_m)]\}}{2(1+v_m)[E_m(1+f_r-2v_m)+2f_m k_r(1-v_m-2v_m^2)]},$$

$$l = \frac{E_m\{v_m f_m[E_m + 2k_r(1+v_m)] + 2f_r k_r(1-v_m^2)\}}{(1+v_m)[E_m(1+f_r-2v_m)+2f_m k_r(1-v_m-2v_m^2)]},$$

$$n = \frac{E_m^2 f_m(1+f_r-f_m v_m) + 2f_m f_r(k_r n_r - l_r^2)(1+v_m)^2(1-2v_m)}{(1+v_m)[E_m(1+f_r-2v_m)+2f_m k_r(1-v_m-2v_m^2)]}$$
$$+ \frac{E_m[2f_m^2 k_r(1-v_m)+f_r n_r(1+f_r-2v_m)-4f_m l_r v_m]}{E_m(1+f_r-2v_m)+2f_m k_r(1-v_m-2v_m^2)},$$

$$p = \frac{E_m[E_m f_m + 2p_r(1+v_m)(1+f_r)]}{2(1+v_m)[E_m(1+f_r)+2f_m p_r(1+v_m)]},$$

$$k = \frac{E_m[E_m f_m + 2m_r(1+v_m)(3+f_r-4v_m)]}{2(1+v_m)\{E_m[f_m+4f_r(1-v_m)]+2f_m m_r(3-v_m-4v_m^2)\}},$$

$$\alpha_r = \frac{3(K_m + G_m) + k_r + l_r}{3(k_r + G_m)}$$

$$\beta_r = \frac{1}{5}\left[\frac{4G_m + 2k_r + l_r}{3(G_m + k_r)} + \frac{4G_m}{G_m + p_r} + \frac{2\left[G_m \frac{(3K_m + G_m)}{} + G_m(3K_m + 7G_m)\right]}{G_m(3K_m + G_m) + m_r(3K_m + 7G_m)}\right]$$

$$\delta_r = \frac{1}{3}\left[n_r + 2l_r + \frac{(2k_r + l)(3K_m + 2G_m - l_r)}{G_m + k_r}\right]$$

$$\eta_r = \frac{1}{5}\left[\frac{2}{3}(n_r - l_r) + \frac{8G_m p_r}{G_m + p_r} + \frac{2(k_r - l_r)(2G_m + l_r)}{3(G_m + k_r)}\right]$$
$$+ \frac{1}{5}\left[\frac{8m_r G_m(3K_m + 4G_m)}{3K_m(m_r + G_m) + G_m(7m_r + G_m)}\right].$$

References

Bellman R, Casti J (1971) Differential quadrature and long term integration. J Math Anal Appl 34:235–238

Bert CW, Malik M (1997) Differential quadrature method: a powerful new technique for analysis of composite structures. Compos Struct 39:179–189

Chern YC, Chao CC (2000) Comparison of natural frequencies of laminates by 3D theory, part II: curved panels. J Sound Vib 230:1009–1030

Cho JR, Tinsley oden J (2000) Functionally graded material: a parameter study on thermal-stress characteristics using the Crank-Nicolson-Galerkin scheme. Comput Meth Appl Mech Eng 188:17–38

Chunyu L, George J, Zhuping D (2001) Dynamic behavior of a cylindrical crack in a functionally graded interlayer under torsional loading. Int J Solids Struct 38:773–785

Dai H (2002) Carbon nanotubes: opportunities and challenges. Surf Sci 500:218–241

Esawi AMK, Farag MM (2007) Carbon nanotube reinforced composites: potential and current challenges. Mater Des 28:2394–2401

Eshelby JD (1957) The determination of the elastic field of an ellipsoidal inclusion, and related problems. Proc R Soc London Ser A 241:376–396

Farid M, Zahedinejad P, Malekzadeh P (2010) Three-dimensional temperature dependent free vibration analysis of functionally graded material curved panels resting on two-parameter elastic foundation using a hybrid semi-analytic, differential quadrature method. Mater Des 31:2–13

Fidelus JD, Wiesel E, Gojny FH, Schulte K, Wagner HD (2005) Thermo-mechanical properties of randomly oriented carbon/epoxy nanocomposites. Compos Part A 36:1555–1561

FGM Forum (1991) Survey for application of FGM, Tokyo, Japan: The Society of Non Tradition Technology

Gojny FH, Wichmann MHG, Fiedler B, Schulte K (2005) Influence of different carbon nanotubes on the mechanical properties of epoxy matrix composites-a comparative study. Compos Sci Technol 65:2300–2313

Han Y, Elliott J (2007) Molecular dynamics simulations of the elastic properties of polymer/carbon nanotube composites. Comput Mater Sci 39:315–323

Han X, Liu GR, Xi ZC, Lam KY (2001) Transient waves in a functionally graded cylinder. Int J Solids Struct 38:3021–3037

Kang I, Heung Y, Kim J, Lee J, Gollapudi R, Subramaniam S (2006) Introduction to carbon nanotube and nanofiber smart materials. Compos B 37:382–394

Lanhe WU (2004) Thermal buckling of a simply supported moderately thick rectangular FGM plate. Compos Struct 64:211–218

Lau KT, Gu C, Hui D (2006) A critical review on nanotube and nanotube/nanoclay related polymer composite materials. Compos B 37:425–436

Loy CT, Lam KY, Reddy JN (1999) Vibration of functionally graded cylindrical shells. Int J Mech Sci 41:309–324

Manchado MAL, Valentini L, Biagiotti J, Kenny JM (2005) Thermal and mechanical properties of single-walled carbon nanotubes-polypropylene composites prepared by melt processing. Carbon 43:1499–1505

Matsunaga H (2008) Free vibration and stability of functionally graded shallow shells according to a 2-D higher-order deformation theory. Compos Struct 84:132–146

Matsunaga H (2009) Free vibration and stability of functionally graded circular cylindrical shells according to a 2D higher-order deformation theory. Compos Struct 88:519–531

Mokashi VV, Qian D, Liu YJ (2007) A study on the tensile response and fracture in carbon nanotube-based composites using molecular mechanics. Compos Sci Technol 67:530–540

Mura T (1982) Micromechanics of defects in solids. Martinus Nijhoff, The Hague

Ng TY, Lam KY, Liew KM, Reddy JN (2001) Dynamic stability analysis of functionally graded cylindrical shells under periodic axial loading. Int J Solids Struct 38:1295–1309

Odegard GM, Gates TS, Wise KE, Park C, Siochi EJ (2003) Constitutive modeling of nanotube reinforced polymer composites. Compos Sci Technol 63:1671–1687

Pradhan SC, Loy CT, Lam KY, Reddy JN (2000) Vibration characteristics of functionally graded cylindrical shells under various boundary conditions. Appl Acoust 61:119–129

Pradyumna S, Bandyopadhyay JN (2008) Free vibration analysis of functionally graded curved panels using a higher-order finite element formulation. J Sound Vib 318:176–192

Seidel GD, Lagoudas DC (2006) Micromechanical analysis of the effective elastic properties of carbon nanotube reinforced composites. Mech Mater 38:884–907

Shen HS (2009) Nonlinear bending of functionally graded carbon nanotube-reinforced composite plates in thermal environments. Compos Struct 91:9–19

Shi DL, Feng XQ, Huang YY, Hwang KC, Gao H (2004) The effect of nanotube waviness and agglomeration on the elastic property of carbon nanotube reinforced composites. J Eng Mat Tech 126:250–257

Shokrieh MM, Rafiee R (2010a) Investigation of nanotube length effect on the reinforcement efficiency in carbon nanotube based composites. Compos Struct 92:2415–2420

Shokrieh M, Rafiee R (2010b) On the tensile behavior of an embedded carbon nanotube in polymer matrix with nonbonded interphase region. Compo Struc 92:647–652

Shu C, Wang CM (1999) Treatment of mixed and non-uniform boundary conditions in GDQ vibration analysis of rectangular plates. Eng Struct 21:125–134

Tahouneh V (2014) Free vibration analysis of thick CGFR annular sector plates resting on elastic foundations. Struct Eng Mech 50:773–796

Tahouneh V, Naei MH (2014) A novel 2-D six-parameter power-law distribution for three-dimensional dynamic analysis of thick multi-directional functionally graded rectangular plates resting on a two-parameter elastic foundation. Meccanica 49:91–109

Tahouneh V, Yas MH (2012) 3-D free vibration analysis of thick functionally graded annular sector plates on Pasternak elastic foundation via 2-D differential quadrature method. Acta Mech 223:1879–1897

Tahouneh V, Yas MH, Tourang H, Kabirian M (2013) Semi-analytical solution for three-dimensional vibration of thick continuous grading fiber reinforced (CGFR) annular plates on

Pasternak elastic foundations with arbitrary boundary conditions on their circular edges. Meccanica 48:1313–1336

Thostenson ET, Ren ZF, Chou TW (2001) Advances in the science and technology of carbon nanotubes and their composites: a review. Compos Sci Technol 61:1899–1912

Tsai SW, Hoa CV, Gay D (2003) Composite materials, design and applications. CRC Press, Boca Raton

Weissenbek E, Pettermann HE, Suresh S (1997) Elasto-plastic deformation of compositionally graded metal-ceramic composites. Acta Mater 45:3401–3417

Yang J, Shen H (2003) Free vibration and parametric resonance of shear deformable functionally graded cylindrical panels. J Sound Vib 261:871–893

Yas MH, Tahouneh V (2012) 3-D free vibration analysis of thick functionally graded annular plates on Pasternak elastic foundation via differential quadrature method (DQM). Acta Mech 223:43–62

Zhu R, Pan E, Roy AK (2007) Molecular dynamics study of the stress–strain behavior of carbon-nanotube reinforced Epon 862 composites. Mater Sci Eng 447:51–57

5

Analysis, prediction, and case studies of early-age cracking in bridge decks

Adel ElSafty[1] · Matthew K. Graeff[2] · Georges El-Gharib[3] · Ahmed Abdel-Mohti[4] ·
N. Mike Jackson[5]

Abstract Early-age cracking can adversely affect strength, serviceability, and durability of concrete bridge decks. Early age is defined as the period after final setting, during which concrete properties change rapidly. Many factors can cause early-age bridge deck cracking including temperature change, hydration, plastic shrinkage, autogenous shrinkage, and drying shrinkage. The cracking may also increase the effect of freeze and thaw cycles and may lead to corrosion of reinforcement. This research paper presents an analysis of causes and factors affecting early-age cracking. It also provides a tool developed to predict the likelihood and initiation of early-age cracking of concrete bridge decks. Understanding the concrete properties is essential so that the developed tool can accurately model the mechanisms contributing to the cracking of concrete bridge decks. The user interface of the implemented computer Excel program enables the user to input the properties of the concrete being monitored. The research study and the developed spreadsheet were used to comprehensively investigate the issue of concrete deck cracking. The spreadsheet is designed to be a user-friendly calculation tool for concrete mixture proportioning, temperature prediction, thermal analysis, and tensile cracking prediction. The study also provides review and makes recommendations on the deck cracking based mainly on the Florida Department of Transportation specifications and Structures Design Guidelines, and Bridge Design Manuals of other states. The results were also compared with that of other commercially available software programs that predict early-age cracking in concrete slabs, concrete pavement, and reinforced concrete bridge decks. The outcome of this study can identify a set of recommendations to limit the deck cracking problem and maintain a longer service life of bridges.

Keywords Cracking · Early-age · Bridge deck · Temperature · Shrinkage

Introduction

Transverse cracking has been observed in many bridge decks in Florida and other states (e.g., Wan et al. 2010). Transverse deck cracking is more likely to occur in early ages. The ACI Committee 231, Properties of Concrete at Early Ages, identified "early age" as the period after final setting (ACI 231 2010). During this period, concrete properties change rapidly. Early-age volume changes are induced by temperature change, hydration, and drying shrinkage. This volume change can lead to early-age cracking due to restraint of volume changes associated with thermal deformation, shrinkage due to hydration reactions, and shrinkage due to drying. Such cracking can adversely affect strength, serviceability, and durability of bridge decks. Also, the development of deck cracking increases the effect of freeze and thaw cycles which may lead to spalling of concrete, and thus, resulting in corrosion of

✉ Ahmed Abdel-Mohti
a-abdel-mohti@onu.edu

[1] Civil Engineering Department, University of North Florida, Jacksonville, FL, USA

[2] Segars Engineering , 1200 Five Springs Rd., Charlottesville, VA, USA

[3] Johnson, Mirmiran & Thompson, Inc., Lake Mary, FL, USA

[4] Civil Engineering Department, Ohio Northern University, Ada, OH, USA

[5] Department of Civil Engineering and Construction Management, Georgia Southern University, Statesboro, GA, USA

steel reinforcement. Transverse deck cracking may also increase carbonation and chloride penetration leading to accelerated steel reinforcement corrosion. Also, a possible damage to underlying components may take place, and the bridge may experience premature deterioration. Therefore, transverse deck cracks affect bridges causing loss of stiffness, and eventually, loss of function, undesirable esthetic condition, reduction of service life of structures, and increase in maintenance costs.

Several studies investigated the issue of deck cracking (Manafpour et al. 2015; Wright et al. 2014; Maggenti et al. 2013; Peyton et al. 2012; Darwin et al. 2012; Slatnick et al. 2011; McLeod et al. 2009; Wan et al. 2010; French et al. 1999; Babaei and purvis 1996; Krauss and Rogalla 1996; La Fraugh and Perenchio 1989; Babaei and Hawkins 1987; PCA 1970). Numerous factors can affect transverse deck cracking in highway bridges including time-dependent material properties, restraints, casting sequence, formwork, and environmental factors. The aforementioned studies determined that span continuity, concrete strength, and girder type are the most important design factors influencing transverse cracking. The design factors most related to transverse cracking are longitudinal restraint, deck thickness, and top transverse bar size. Material properties such as cement content, cement composition, early-age elastic modulus, creep, aggregate type and quantity, heat of hydration, air content, and drying shrinkage also influence deck cracking. Schmitt and Darwin (1999) conducted a study considering various site conditions factors such as average air temperature, low air temperature, high air temperature, daily temperature range, relative humidity, average wind velocity, and evaporation. To investigate the concrete cracking in new bridge decks and overlays, Wan et al. (2010) completed a Wisconsin Department of Transportation (DOT) project indicating that the rapid development of compressive strength and modulus of elasticity of concrete may lead to significant shrinkage and tensile stresses in the deck. Table 1 shows a sample of departments of transportation projects addressing concrete bridge deck cracking.

Research objectives

The objective of this research study is to investigate the early-age cracking and its mitigation. In this study, a tool was developed to facilitate predicting the early-age cracking in bridge decks. It helps in predicting cracks in both under-construction and future bridge decks. Two case studies were investigated to examine the issue of cracking and to compare the outcome of the developed tool to that of an available software program "HIPERPAV". After verification, the tool was used to check the cracking tendency in newer bridges with different material properties. Also, field investigation was conducted to observe deck cracking in existing bridge decks and monitor the development of transverse cracks in new bridge under construction. In general, the ultimate goal of this research study is to provide a comprehensive insight of the issue of early-age cracking in bridge deck and to provide recommendations to limit the problem.

Field investigation

Field investigation was conducted in this study to observe the pattern and locations of cracks developed in several existing bridge decks (Fig. 1a–h). A number of bridges were inspected, assessed, and repaired with sealers including Florida bridges in Fort Lauderdale, Jacksonville, and Pensacola. Investigation was also performed to observe the crack development in a new bridge "US 1 Bridge" under construction. Most of the observed cracks have an average width of 0.02 inch, and they are spaced about 3–4 ft apart. A number of tests were performed in a previous study to evaluate and treat existing cracks (ElSafty and Jackson 2012; ElSafty and Abdel-Mohti 2013; El Safty et al. 2013). The results were in agreement with Manafpour et al. (2015) in which it was observed that initial observation of early-age deck cracks was within the first 2 months after concrete placement and was more pronounced during summer time.

Table 1 DOTs concrete bridge deck cracking

Department of transportation	Year	Title
PennDOT	2015	Bridge deck cracking: effects on in-service performance, prevention, and remediation
FDOT	2012	Sealing of cracks on florida bridge decks with steel girders
NYDOT	2011	Tool for analysis of earlyage transverse cracking of composite bridge decks
WisconsinDOT	2010	Concrete cracking in new bridge decks and overlays
ALDOT	2010	Evaluation of cracking of the US 331 bridge deck
CDOT	2003	Assessment of the cracking problem in newly constructed bridge decks in Colorado
MDOT	1998	Transverse cracking in bridge decks: field study

Fig. 1 Bridge deck cracking.
a US 1 bridge (Jacksonville, Florida), **b** Close up view of deck crack, **c** JTB bridge (Jacksonville, Florida), **d** Deck crack pattern, **e** Fort Lauderdale bridge (Florida), **f** Close up view of the deck crack, **g** Blackwater River Bridge (Milton-Pensacola, Florida), **h** Core sampling showing crack development over the transverse reinforcement

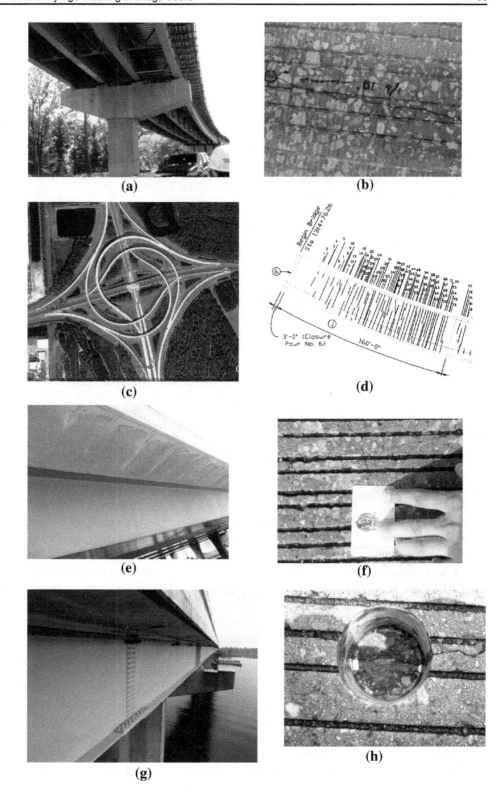

(a)   (b)

(c)   (d)

(e)   (f)

(g)   (h)

Development of a tool to predict deck cracking

Understanding thoroughly the concrete properties is vital to accurately model the mechanisms contributing to the cracking of concrete decks. The deck cracking Excel spreadsheet tool was developed in this study to facilitate conducting a number of case studies and to predict the potential transverse deck cracking of bridges. The user interface of the developed Excel program enables the user to input the properties of the concrete being monitored. The

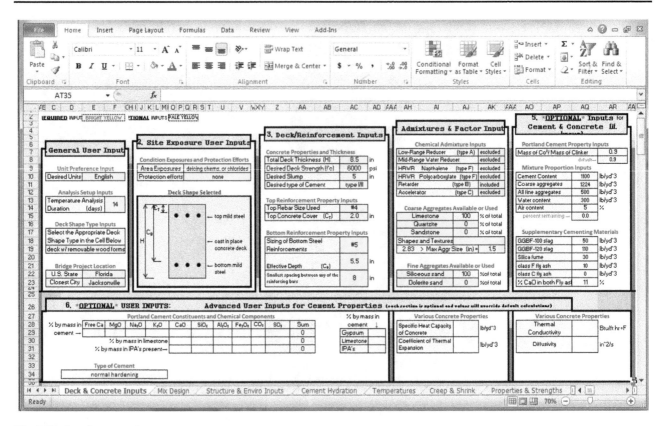

Fig. 2 Deck and concrete inputs

spreadsheet is designed to be a user-friendly calculation tool for concrete mixture proportioning, temperature prediction, thermal analysis, and tensile cracking prediction. It is designed specifically for concrete bridge decks. It also addresses different types of construction approaches including a deck with a stay-in-place galvanized metal pan, a deck with removable forms, and a deck on a precast panel. The user should address the fundamental principles and mechanics of concrete hardening to obtain accurate temperatures, thermal stresses, and cracking risk calculations. Also, it is advisable to follow available recommendations to limit concrete deck cracking such as Manafpour et al. (2015). The aspects of concrete hardening addressed in the spreadsheet is subdivided into multiple sections; the first being concrete mixture proportioning, followed by temperature prediction, thermal stress analysis, and finally tensile cracking predictions.

The spreadsheet has multiple tabs that show the logical steps that a designer would follow during a design process. The tabs include Deck and Concrete Inputs, Mix Design, Structural and Environmental Inputs, Cement Hydration, Temperature Analysis, Properties and Strengths, Creep and Shrinkage Stresses, and Result Summary. Each tab may contain various required user inputs, optional user inputs, default values, or calculated values. Each of the cells is color coded. Examples of a user input tab is shown in Fig. 2.

Recommendations for design inputs

A designer would need to determine all the necessary design inputs to achieve an accurate assessment of the likelihood of reinforced concrete deck cracking. Figure 2 summarizes all the necessary deck and concrete inputs. The Deck and Concrete Inputs worksheet values are in accordance with the current engineering practices. Since the Materials Research Report which is titled Sealing of Cracks on Florida Bridge Decks with Steel Girders, was created for FDOT, it is recommended to verify the input values against the most recent FDOT Standard Specifications for Road and Bridge Construction (FDOT Specs) and the FDOT Structures Manual Volume 1 Structures Design Guidelines (FDOT SDG). Other Departments of Transportation Bridge Design Manuals were also used (i.e., Ohio DOT, Pennsylvania DOT, Indiana DOT).

The summary below is a list of recommendations for the Deck/Reinforcement Inputs:

1. Total deck thickness (H): The FDOT SDG requires that cast-in-place deck thickness shall be 8.0 inches for short bridges and 8.5 inches for long bridges (SDG 4.2.2). The determining length is the length of the bridge structure measured along the profile grade line (PGL) from front face of backwall at Begin Bridge to

Table 2 Range of user override water adjustment factors

Factor	Adjustment ranges (negative is reduction)			
Normal range water reducer (ASTM type A)	0	%	−10	%
Mid-range water reducer	−8	%	−15	%
High range water reducer (ASTM type F)	−12	%	−30	%
Air entrainment effect	5	lbs/% air needed for desired %		
Aggregate shape and texture	−20	lbs	−45	lbs
Aggregate gradation	10	%	−10	%
Supplementary mineral admixtures	15	%	−10	%
Other unspecified factors	10	%	−10	%

front face of backwall at End Bridge of the structure. Short bridges and long bridges are defined in Section 4.2.1 of the FDOT SDG as follows:

(a) Short bridges: bridge structures less than or equal to 100 ft in PGL length.

(b) Long bridges: bridge structures more than 100 ft in PGL length.

The Ohio DOT (ODOT) Bridge Design Manual (BDM) requires calculating a minimum reinforced concrete deck thickness using the following equation (BDM 302.2.1):

$$T_{\min}(\text{inches}) = (S + 17)(12)/36 \geq 8.5 \text{ inches} \qquad (1)$$

where S is the effective span length in feet determined in accordance with AASHTO LRFD 9.7.3.2 (AASHTO 2007). The minimum deck thickness includes a 1 inch wearing surface.

ODOT also offers a Concrete Deck Design Aid table that shows the deck thickness based on the effective span length. The deck thicknesses in this table range from 8.5 to 10.5 inches. ODOT does not allow the use of precast deck panels, since they have shown cracking problems at the joints between the panels, and there are questions on the transfer of stresses in the finished deck sections. The Pennsylvania DOT (PennDOT) Design Manual Part 4 Structures (DM-4) recommends a minimum reinforced concrete deck thickness of 8.0 inches (Part B, 9.7.1.1) which includes a 0.5 inch wearing surface. This is also true for both reinforced and prestressed precast concrete deck panels (Part B, 9.7.5.1). The Indiana DOT (INDOT) Design Manual Chapter 404 (Ch404) Bridge Deck requires the reinforced deck depth to be a minimum of 8.0 inches that includes 0.5 inch of sacrificial wearing surface (404-2.01.2). The AASHTO LRFD Bridge Design Specifications (LRFD) requires that the minimum thickness of a reinforced concrete deck should not be less than 7.0 inches (LRFD 9.7.1.1), if approved by the owner. This minimum thickness does not include any provision for grinding, grooving, and sacrificial surface. For precast decks on girders, the minimum thickness of a reinforced concrete deck and prestressed concrete deck should not be less than 7.0 inches (LRFD 9.7.5.1). Based on the information

mentioned above, it is recommended that a range of 7.0–11.0 inches with 0.5 inch increments be used in the total deck thickness.

2. Desired deck strength ($f'c$): The FDOT SDG Table 1.4.3-1 requires that cast-in-place concrete deck shall have the following structural class:

   (a) Class II (bridge deck) for slightly aggressive environment.

   (b) Class IV for moderately and extremely aggressive environment.

   The environmental classification is based on Section 1.3 of the SDG. Concrete classes are defined in Section 346, Portland cement concrete, of the FDOT Specifications. The concrete specified minimum strength at 28 days is shown in the list below:

   (a) For class II concrete (bridge deck): 4500 psi.

   (b) For class IV concrete: 5500 psi.

   The ODOT BDM requires the deck concrete to be Class S or Class HP (BDM 302.1.2.2) with a minimum 28-day compressive strength of 4500 psi (BDM 302.1.1). The PennDOT DM4 requires the deck concrete to be Class AAA with a minimum 28-day compressive strength of 4000 psi (DM4 Part B, 5.4.2.1). The INDOT Ch404 requires the deck concrete to be Class C (404-2.01.6) with a minimum 28-day compressive strength of 4000 psi (404-2.01.7). Based on the information mentioned above, this research considered a range of 3000–5500 psi.

3. Desired Slump: The FDOT Specifications Section 346-3.1. Table 2 shows a target slump value of 3 inches for both Class II (Bridge Deck) and Class IV concrete with a ±1.5 inch tolerance (Specs 346-6.4). ACI 301 requires a slump of 4 inches at the point of delivery (ACI 301-4.2.2.2) with a tolerance of ±1 inch (ACI 301 2010). Based on the information mentioned above, the research considered a range of 1–7 inches in the desired slump.

4. Top concrete cover: The FDOT SDG requires that cast-in-place decks shall have the following top concrete cover (Table 1.4.2-1):

(a)   Short bridges: 2 inches.

(b)   Long bridges: 2.5 inches.

The ODOT BDM requires a 2.5 inches minimum cover for the concrete deck top surface (BDM 301.5.7). The PennDOT DM4 requires a 2.5 inches minimum cover for the concrete deck top surface (DM4 Part B, 5.12.3). The INDOT Ch404 requires a 2.5 inches minimum cover for the concrete deck top surface (404-2.01.3). Based on the information mentioned above, this research considered a range of 2–3.0 inches to be used in the top concrete cover.

5.   Effective depth: The FDOT SDG requires that cast-in-place decks shall have a bottom reinforcement concrete cover of 2.0 inches (Table 1.4.2-1). The ODOT BDM requires a 1.5 inch minimum cover for the concrete deck bottom surface (BDM 301.5.7). The PennDOT DM4 requires a 1.0 inch minimum cover for the concrete deck bottom surface (DM4 Part B, 5.12.3). The INDOT Ch404 requires a 1.0 inch minimum cover for the concrete deck bottom surface (404-2.01.3). The effective depth is a function of the deck thickness and the reinforcement cover to the bottom surface of deck. Assuming that the deck thickness varies from 7 to 11 inches and the minimum cover to the bottom surface is 1.0 inch, this research considered a range of 6–10.0 inches to be used in the effective depth.

Mix design worksheet

The primary source commonly used for the mixture proportioning calculations is the ACI 211.1-91 document "Standard Practice for Selecting Proportions for Normal, Heavyweight, and Mass Concrete" (ACI 211-91 2009). The basic steps are as follows:

1.   Determine the amount of water needed to achieve a given slump for the maximum aggregate size selected by the designer and to make the required adjustments to the water content based on the material properties, chemical admixtures, and entrained air properties.

2.   Determine the water to cement ratio needed to achieve a desired strength with the percent of entrained air specified, where the use of supplementary materials is assumed to not affect the water to cement ratio needed to achieve the desired strength.

3.   Calculate the coarse aggregate fraction based on the maximum size of aggregate selected and the fine aggregate fineness modulus.

4.   Calculate the required amount of fine aggregates to fill the remaining concrete volume since the volume of the cementitious materials, the water content, the coarse

aggregate content, and the percentage of air are already included. The fine aggregate weight is then calculated from the volume using the specific gravity of the sands.

In this research, the calculations for concrete mixture proportioning, following the steps presented above, are only performed if the user does not specify a predetermined mixture (Fig. 3). The water content can be adjusted for several factors both with and without a user defined concrete mixture proportion. Table 2 presents typical water adjustment factors.

Structure and environmental inputs

The structure and environmental design inputs are important and will affect the possibility of cracking in bridge decks. The designer will need to determine the properties of both of the deck and girders. Also, information such as type and duration of curing, time of placing concrete, method of curing, temperature at casting, age of concrete at loading, and properties of formwork used, and predication of weather condition after casting are important. Figure 4 shows an example of design inputs.

Temperature prediction

Since early-age properties of concrete change rapidly, the thermal properties of the concrete and its constituents shall be updated at each given time. Some of the time-dependent properties include: thermal conductivity and the specific heat of the concrete. These properties shall be calculated at each time of interest.

Concrete thermal properties

(a) *Thermal Conductivity*: The thermal conductivity is known to be a function of "the moisture content, content and type of aggregate, porosity, density and temperature (Van Breugel 1998)." The concrete thermal conductivity increases with increasing moisture content. Based on the recommendation of Schindler (2002), this research assumes a linear decrease of the thermal conductivity with the degree of hydration from 1.33 times the ultimate thermal conductivity to the ultimate thermal conductivity as shown in Eq. 2:

$$k_c(\alpha) = k_{uc} \cdot (1.33 - 0.33 \cdot \alpha) \qquad (2)$$

where k_c is the concrete thermal conductivity (W/m/K), α is the degree of hydration, and k_{uc} is the ultimate hardened concrete thermal conductivity.

(b) *Specific heat capacity*: The specific heat of concrete is also dependent on the mixture proportions, the degree of hydration, moisture levels, and the temperature (Schindler

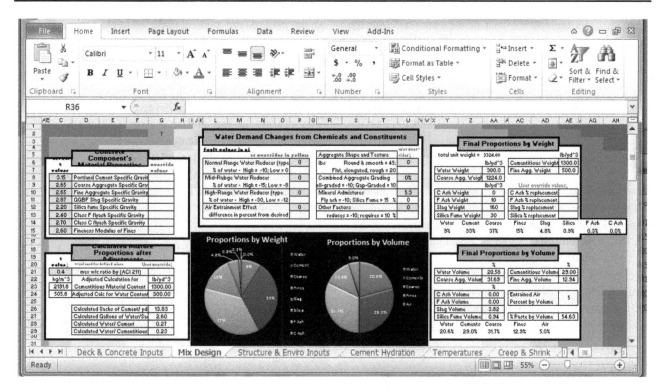

Fig. 3 Mix design

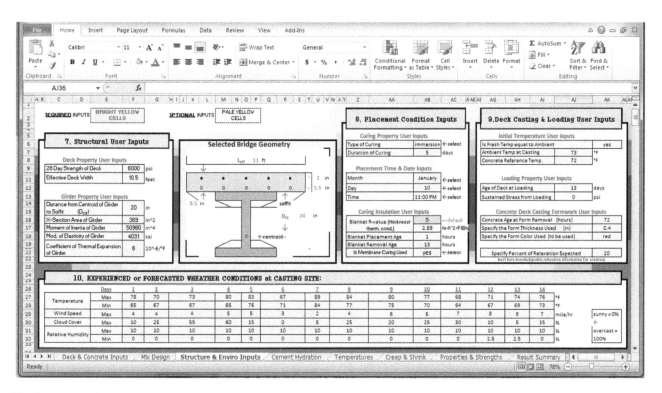

Fig. 4 Example of structural and environmental inputs

2002). A model proposed by Van Breugel accounts for changes in the specific heat based on degree of hydration, mixture proportions, and temperature as shown in Eq. 3.

$$c_{pconc} = \frac{1}{\rho_{conc}} \cdot (W_c \cdot \alpha \cdot c_{ref} + W_c \cdot (1 - \alpha) \cdot c_c + W_a \cdot c_a + W_w \cdot c_w) \quad (3)$$

where c_{pconc} is the specific heat of the concrete (J/kg/K), ρ_{conc} is the concrete density (kg/m^3), W_c is the weight of cement (kg/m^3), W_a is the weight of aggregate (kg/m^3), W_w is the weight of water (kg/m^3), c_c is the cement specific heat (J/kg/K), c_a is the aggregate specific heat (J/kg/K), c_w is the water specific heat (J/kg/k), and c_{ref} is an average ultimate specific heat of the cement taken as 840 (J/kg/K).

Concrete heat of hydration

This research accounts for the change in thermal properties of the concrete and its constituents by updating for the change at any given time. For temperature due to hydration, the concrete mix design is first modified using the Bogue calculations according to ASTM C 150. The concrete heat of hydration parameters H_u, τ, β, α_u, and E_a are then calculated based on the concrete mixture proportions and the constituent material properties. The τ, β, and α_u parameters are calculated from Eqs. 4, 5, and 6.

$$\alpha_u = \frac{1.031 \cdot w/cm}{0.194 + w/cm}$$
$$+ \exp\left\{\begin{array}{l} -0.885 - 13.7 \cdot p_{C_4AF} \cdot p_{cem} \\ -283 \cdot p_{Na_2O_{eq}} \cdot p_{cem} \\ -9.9 \cdot p_{FA} \cdot p_{FA-CaO} \\ -339 \cdot WRRET - 95.4 \cdot PCHRWR \end{array}\right\} \quad (4)$$

$$\tau = \exp$$
$$\times \left\{\begin{array}{l} 2.68 - 0.386 \cdot p_{C_3S} \cdot p_{cem} + 105 \cdot p_{Na_2O} \cdot p_{cem} + 1.75 \cdot p_{GGBF} \\ -5.33 \cdot p_{FA} \cdot p_{FA-CaO} - 12.6 \cdot ACCL + 97.3 \cdot WRRET \end{array}\right\}$$
$$(5)$$

$$\beta = \exp$$
$$\times \left\{\begin{array}{l} -0.494 - 3.08 \cdot p_{C_3A} \cdot p_{cem} - 0.864 \cdot p_{GGBF} \\ +96.8 \cdot WRRET + 39.4 \cdot LRWR + 23.2 \cdot MRWR \\ +38.3 \cdot PCHRWR + 9.07 \cdot NHRWR \end{array}\right\}$$
$$(6)$$

Similarly, the parameters of heat and activation energy are also calculated based on the concrete mixture proportions and the constituent material properties as described by Eqs. 7, 8, and 9.

$$Hu = \left\{\begin{array}{l} H_{cem} \cdot p_{cem} + 461 \cdot p_{GGBF-100} + 550 \cdot p_{GGBF-120} \\ +1800 \cdot p_{FA-CaO} \cdot p_{FA} + 330 \cdot p_{S.F.} \end{array}\right\}$$
$$(7)$$

$$H_{cem} = \left\{\begin{array}{l} 500 \cdot p_{C_3S} + 260 \cdot p_{C_2S} + 866 \cdot p_{C_3A} + 420 \cdot p_{C_4AF} \\ +624 \cdot p_{SO_3} + 1186 \cdot p_{freeCa} + 850 \cdot p_{MgO} \end{array}\right\}$$
$$(8)$$

$$E_a = \left\{\begin{array}{l} 41230 + 8330 \cdot [(C_3A + C_4AF) \cdot p_{cem} \cdot Gypsum \cdot p_{cem}] \\ -3470 \cdot Na_2O_{eq} - 19.8 \cdot Blaine + 2.96 \cdot p_{FA} \cdot p_{CaO-FA} \\ +162 \cdot p_{GGBFS} - 516 \cdot p_{S.F.} - 30900 \cdot WRRET - 1450 \cdot ACCL \end{array}\right\}$$
$$(9)$$

where p_{C3S} is the percent alite content in the portland cement, p_{C3A} is the percent aluminate in the portland cement, p_{C2S} is the percent belite in the portland cement, p_{C4AF} is the percent ferrite in the portland cement, p_{SO3} is the percent total sulfate in the portland cement, p_{MgO} is the percent MgO in the portland cement, and p_{freeCa} is the percent CaO in the portland cement. Table 3 presents typical dosages of chemical admixtures.

The maturity method used to determine the rate of hydration of the cement is the equivalent age method described in ASTM C 1074 where the equivalent age of the concrete is calculated as described in Eq. 10.

$$t_e = \sum e^{-\frac{E_a}{R} \cdot \left(\frac{1}{(T_a+273)} - \frac{1}{(T_r+273)}\right)} \Delta T \quad (10)$$

The degree of hydration is next calculated by use of Eq. 8, and ultimately, the rate of heat generated is calculated using the parameter values from Eqs. 4 through 11 at any given time using Eq. 12 (Schindler and Folliard 2005).

$$\alpha(t_e) = \alpha_u \cdot \exp\left(-\left[\frac{\tau}{t_e}\right]^\beta\right) \quad (11)$$

$$Q(t) = H_u \cdot C_c \cdot \left(\frac{\tau}{t_e}\right)^\beta \cdot \left(\frac{\beta}{t_e}\right) \cdot \alpha_u \cdot \exp\left(-\left[\frac{\tau}{t_e}\right]^\beta\right)$$
$$\cdot \exp\left(\frac{E_a}{R}\left(\frac{1}{273 + T_r} - \frac{1}{273 + T}\right)\right) \cdot \left(\frac{1}{3600}\right)$$
$$(12)$$

where t_e is the concrete equivalent age at the reference temperature as shown in Eq. 12 (h), H_u is the total amount of heat generated at 100 % hydration (J/kg), C_c is the total amount of cementitious materials (kg/m^3), τ is the hydration time parameter (h), β is the hydration slope parameter, α_u is the ultimate degree of hydration, E_a is the activation energy (J/mol), R is the universal gas constant (J/mol/K), T_r is the reference temperature (°C), and T_a is the average temperature during the time interval. At this point the degree of hydration, concrete maturity, rate of heat generation, and the adiabatic temperature rise can be calculated. Figure 5 shows an example of the previous

Table 3 Default chemical admixture dosages assumed if selected but not specified

Chemical admixture	Default percent used if not specified	
LRWR	0.0029	% By mass of cementitious materials
MRWR	0.0032	% By mass of cementitious materials
WRRET	0.0035	% By mass of cementitious materials
NHRWR	0.0078	% By mass of cementitious materials
PCHRWR	0.0068	% By mass of cementitious materials
ACCL	0.013	% By mass of cementitious materials

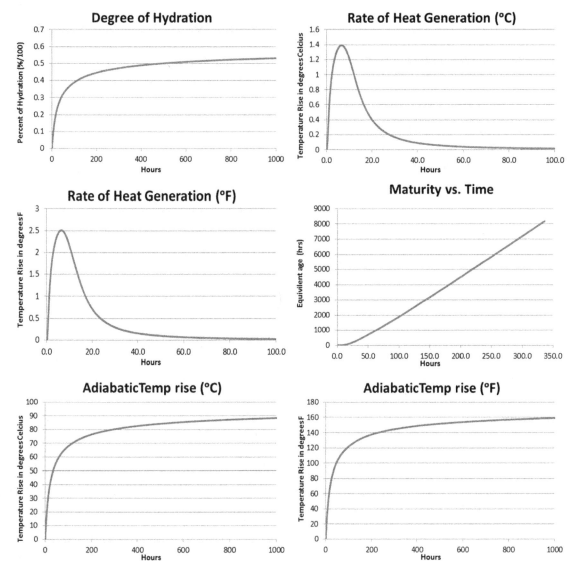

Fig. 5 Example of hydration properties

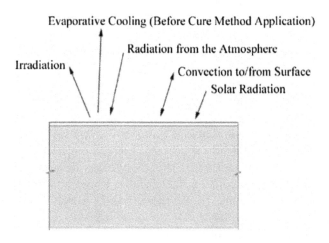

Fig. 6 Boundary conditions used for temperature analysis model

calculations which displays the graphs generated in this research.

Boundary conditions

The boundary conditions considered for the temperature analysis include many heat sources and sinks. The primary conditions models include: evaporative cooling, free and forced convection, conduction, atmospheric radiation, solar radiation and irradiation. A depiction of the boundary conditions modeled is shown in Fig. 6.

Evaporative cooling

The evaporative cooling model is from Schindler (2002). The model is reportedly based on the work of Menzel that applied water evaporation rate equations developed by Koehler to concrete. The evaporation rate follows Dalton's law, which relates the water–vapor pressure of the air, at the water surface, and the wind speed to the evaporation rate (Hover 2006). Menzel's equation is shown as Eq. 13 (Al-Fadhala and Hover 2001).

$$E_w = 0.315(e_0 - RH \cdot e_a)(0.253 + 0.060w) \tag{13}$$

where E_w is the water evaporation rate (kg/m^2/h), e_0 is the water surface saturated water vapor pressure (mmHg), e_a is the air water vapor pressure (mmHg), RH is the relative humidity (as a decimal), and w is the wind speed (m/s). The amount of evaporation from concrete may be related to the amount of evaporation from a water surface by Eq. 14 (Schindler 2002):

$$E_c = E_w \cdot \exp\left[-\left(\frac{t}{a_{evap}}\right)^{1.5}\right] \tag{14}$$

where E_c is the evaporation rate from concrete (kg/m^2/h), t is the time from mixing (h), and a_{evap} is mixture-dependent time constant (h). The default value for a_{evap} is equal to 3.75 h, and the evaporative cooling model is applied until either a cure method is applied or 24 h after placing. The final change in heat due to evaporative cooling is calculated using Eq. 15.

$$\Delta Q = -E_c \cdot h_{lat} \tag{15}$$

where ΔQ is the heat lost due to evaporative cooling, E_c is the evaporation rate from the concrete as calculated in Eq. 14, and h_{lat} is calculated by using Eq. 16 where T_{sw} is the temperature of the surface water.

$$h_{lat} = 2,500,000 + 1859 \cdot T_{sw} \tag{16}$$

Convection

Both the free and forced convection heat exchanges are modeled using Eqs. 17 and 18. Equation 17 is defining the change in heat due to the convection process, and Eq. 18 is defining the convection coefficient.

$$\Delta Q = h(T_s - T_a) \tag{17}$$

$$h = C \cdot 0.2782 \cdot \left(\left[\frac{1}{T_{avg} + 17.8}\right]^{0.181}\right) \cdot \left(|T_s - T_a|^{0.266}\right) \cdot \left(\sqrt{1 + (2.8566 \cdot w)}\right) \tag{18}$$

where ΔQ is the change in heat, h is the convection coefficient, T_s is the temperature of the concrete surface, T_a is the temperature of the air, and T_{avg} is the average of the two temperatures.

Conduction

Conduction is the heat lost or gained from the contact of the concrete with any other material or substance. Conduction can be considered to act between the concrete and the air, between the concrete and the form work, or between the concrete and stagnate surface water; and is calculated using Eq. 19.

$$\Delta Q = -k \cdot A \cdot \frac{\Delta T}{\Delta y} \cdot \Delta t \tag{19}$$

where ΔQ is the change in heat, k is the thermal conductivity of the concrete, A is the area of contact, ΔT is the difference in temperature of the two materials, Δy is the thickness of the volume considered, and Δt is the duration of the time interval.

Radiation

The radiation that affects the curing concrete deck occurs as solar radiation, atmospheric radiation, and irradiation. The atmospheric radiation and irradiation are the easiest to calculate, and the respective equations are listed below in Eqs. 20 and 21.

$$\Delta Q = \sigma \cdot \varepsilon_a \cdot T_a^4 \tag{20}$$

$$\Delta Q = \varepsilon_c \cdot \sigma \cdot T_c^4 \tag{21}$$

where ΔQ is the change in heat due to the radiation, σ is the Boltzmann constant (W/m^{2}K^4), ε_a and ε_c are the emissivity values for either the air or the concrete, and T_a and T_c are the temperatures of either the air or the concrete. Solar radiations are much more complicated of calculations requiring calculated values for extraterrestrial radiation, solar declination angles, solar hour angles, and angles of incidence. These values are calculated based on the assigned latitude and longitude of the nearest location selected and follow the procedures outlined in "Solar Engineering of Thermal Processes: Third Edition" by J.A. Duffie and W.A. Beckman. However, the final equation used to calculate the solar radiation on the deck surface at any given time is defined in Eq. 22.

$$\Delta Q = (0.91 - 0.7 \cdot Cc) \cdot G_{on} \cdot Ab_c \tag{22}$$

where ΔQ is the change in heat due to solar radiation, Cc is the percent of cloud cover, G_{on} is the extraterrestrial radiation that would hit the surface, and Ab_c is the absorptiveness of the concrete. Using the values calculated for Eqs. 15, 17, 19, 20, 21, and 22, the final temperature of the concrete accounting for the energy lost or gained is ultimately compiled to generate a graph of temperature versus time. An example of the generated graphs is available in Fig. 7.

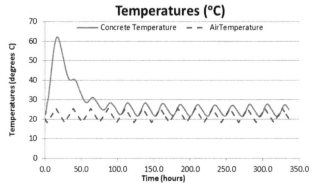

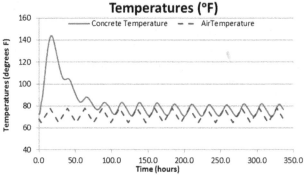

Fig. 7 Depiction of temperature analysis tab, displaying output graphs of temperatures

Thermal stress analysis

Thermal stress modeling in concrete members is nonlinear because of changing early-age material properties such as Poisson's ratio, the coefficient of thermal expansion (CTE), the modulus of elasticity, and the concrete strength. The nonlinearity is also attributed to differential temperature development and creep. The thermal stress analysis includes the evaluation of thermal expansion stresses, shrinkage stresses, the degrees of restraint, and the creep stresses developed over time. The B3 model associated with Zdenek P. Bazant and Sandeep Baweja was the primary source for the creep and shrinkage calculations where thermal, shrinkage, and creep strains are calculated and converted to stresses. The stresses are calculated from the strain values using Eq. 23.

$$\sigma = \frac{E_c \varepsilon}{(1 + v) \cdot (1 - 2v)} \tag{23}$$

where σ is the developed stress, ε is the previously calculated strain, v is the Poisson's ratio, and E_c is the modulus of elasticity of the concrete.

Concrete mechanical properties

Concrete mechanical property development at early ages is dependent on the concrete degree of hydration and

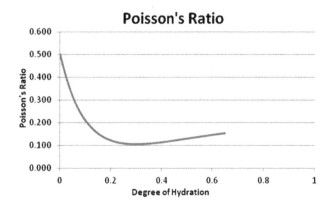

Fig. 8 Graphical depiction of example Poisson's ratio development

temperature development. The mechanical property development is calculated using the equivalent age maturity (ASTM C 1074, 2004) as previously discussed.

(a) *Poisson's ratio*: A multitude of different equations have been developed to relate the maturity to the development of Poisson's ratio. This research uses a proposed model from Deschutter and Taerwe (1996); where Poisson's ratio is based on the degree of hydration as described in Eq. 24.

$$v(\alpha) = 0.18 \cdot \sin\left(\frac{\pi \cdot \alpha}{2}\right) + 0.5 e^{-10\alpha} \tag{24}$$

where $v(\alpha)$ is Poisson's ratio at that degree of hydration and α is the degree of hydration as calculated from the heat of hydration analysis. An example of the graphical display of the Poisson's ratio used in this research is displayed in Fig. 8 (full hydration is not achieved in this example).

(b) *Coefficient of Thermal Expansion*: This research uses a constant CTE because of the lack of a data to model how the mixture proportions relate to CTE development. The constant coefficient of thermal expansion used is calculated from the mixture proportions and the aggregate type using the method proposed by Emanuel and Hulsey (1977) shown in Eq. 25.

$$\alpha_{cteh} = \frac{\alpha_{ca} \cdot V_{ca} + \alpha_{fa} \cdot V_{fa} + \alpha_p \cdot V_p}{V_{ca} + V_{fa} + V_p} \tag{25}$$

where α_{cteh} is the hardened concrete CTE, α_{ca} is the coarse aggregate CTE ($\mu\varepsilon/°C$), V_{ca} is the coarse aggregate volume (kg/m^3), α_{fa} is the fine aggregate CTE ($\mu\varepsilon/°C$), V_{fa} is the fine aggregate volume (kg/m^3), α_p is the paste CTE ($\mu\varepsilon/°C$), and V_p is the paste volume (kg/m^3). The default values of CTE for various constituents presented in Table 4 can be used for evaluation of the concrete's CTE in Eq. 25.

(c) *Compressive Strength*: The compressive strength of the concrete can be calculated in a number of ways. This research calculated the compressive strength of the concrete using two different methods and averages the results. The first method is described by Eq. 26.

Table 4 Default CTE values of concrete constituents used if no modifications are selected by user

Possible concrete constituents	Default CTE values used	
Hardened cement paste	10.8	$\mu\ \varepsilon/°C$
Limestone aggregates	3.5	$\mu\ \varepsilon/°C$
Siliceous river gravel and sands	11	$\mu\ \varepsilon/°C$
Granite aggregates	7.5	$\mu\ \varepsilon/°C$
Dolomitic limestone aggregates	7	$\mu\ \varepsilon/°C$

$$f'c(t) = f'c_{28} \cdot \exp\left(-\left[\frac{\tau_s}{t}\right]^\beta\right) \tag{26}$$

where $f'c(t)$ is the concrete compressive strength at any given time t, $f'c_{28}$ is the concrete compressive strength at 28 days, τ_s is a fit parameter taken as 0.721, and β is another fit parameter taken as 27.8. In the other method, Eq. 27 is solved for $f'c(t)$ using the value from Eq. 28 as $E_c(t)$ and is averaged with the value attained from Eq. 26.

(d) *Modulus of Elasticity*: The elastic modulus provides the correlation between restrained strains and stresses, and it is known to be dependent on the mixture proportions, unit weight, maturity, aggregate modulus, strength, and moisture condition. The elastic modulus is also known to develop faster than the tensile and compressive strengths. In this research, two methods of calculating the modulus are performed and then averaged. The two methods are described by Eqs. 27 and 28 where Eq. 27 is from the ACI 318 document, and Eq. 28 is from the CEB-FIP document.

$$E_c(t) = 57,000\sqrt{f'c(t)} \tag{27}$$

$$E_c(t) = E_{c28}e^{\left[s/2\left(1-\sqrt{28/t}\right)\right]} \tag{28}$$

where $E_c(t)$ is the concrete modulus of elasticity at any time t, E_c28 is the concrete modulus of elasticity at 28 days, and s is a cement type coefficient which is 0.2 for high early strength cements, 0.25 for normal hardening cements, and 0.38 for slow hardening cements.

Thermal expansion

Thermal dilation stresses developed in the concrete are the easiest stresses to calculate using the B3 model. The thermal dilation strain is defined as listed in Eq. 29.

$$\varepsilon_T(t) = \alpha \cdot \Delta T(t) \tag{29}$$

where ε_T is the thermal strain developed at time t, α is the concrete CTE as calculated in Eq. 25, and $\Delta T(t)$ is the difference in temperature from the reference temperature at time t. The relating thermal stresses are then calculated using Eq. 23.

Shrinkage

Concrete early-age free shrinkage strains are dependent on the concrete degree of hydration and temperature development. The free shrinkage strain is composed of the concrete thermal strains, the autogenous strains, the drying shrinkage strains, and the plastic shrinkage strains. In the B3 model, the shrinkage is first estimated from the concrete strength and composition, Eq. 30.

$$\varepsilon_{sh}(t) = -\varepsilon_{sh\infty} \cdot k_h \cdot S(t) \tag{30}$$

where $\varepsilon_{sh}(t)$ is the mean shrinkage strain in the cross section, $\varepsilon_{sh\infty}$ is the time dependence of ultimate shrinkage, k_h is the humidity dependence, and $S(t)$ is the time dependence for shrinkage. These variables can easily be calculated using the B3 model.

Fig. 9 Example of stresses w/o relaxation calculated

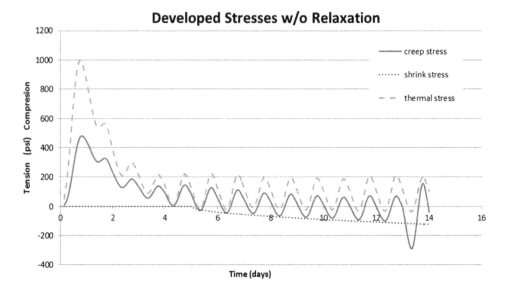

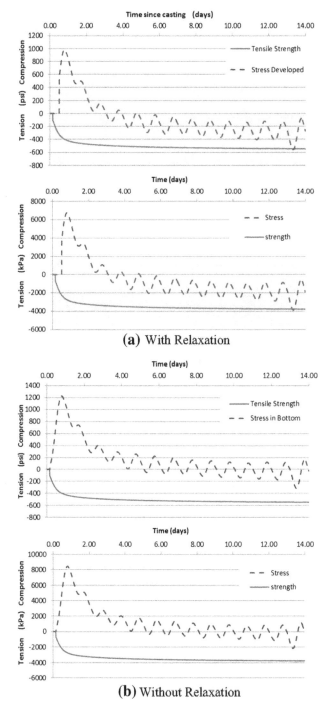

(a) With Relaxation

(b) Without Relaxation

Fig. 10 Stresses and tensile strengths with time. **a** With relaxation, **b** without relaxation

Creep

The creep calculated in this research is primarily due to the applied stresses from early-age thermal stresses and shrinkage stresses prior to loading. The final equation for the calculation of the early-age creep strains is defined in Eq. 31.

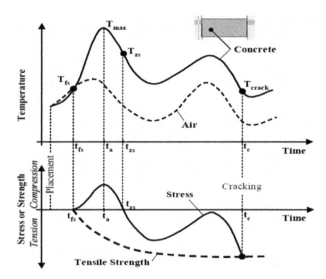

Fig. 11 Documented behavior of hardening concrete and crack identification (Schindler 2002)

$$\varepsilon_{cr}(t) = J(t) \cdot \sigma(t) \tag{31}$$

where $\varepsilon_{cr}(t)$ is the creep strain at any time t, $J(t)$ is the creep compliance function as described in Eq. 32, and $\sigma(t)$ is the stress in the concrete at any time t.

$$J(t) = q_1 + C_0(t) + C_d(t) \tag{32}$$

where $J(t)$ is as previously defined, q_1 is the instantaneous strain due to a unit stress, $C_0(t)$ is the compliance function for basic creep at any time t, and $C_d(t)$ is the compliance function for additional creep due to simultaneous drying. The aforementioned compliance functions can also be easily calculated following the B3 model for creep and shrinkage. An example of the developed stresses calculated is available in Fig. 9.

Tensile cracking prediction

Degrees of restraint

For the degree of restraint, the restraining materials modulus is defined as E_f, and the modulus of the freshly casted concrete is E_c. The ratio of the two moduli defines the degree of restraint as described in Eq. 33.

$$K_r = \begin{cases} 0.2 & \text{when} & \dfrac{E_f}{E_c} \leq .1 \\[2mm] .33 & \text{when} & .1 < \dfrac{E_f}{E_c} \leq .2 \\[2mm] .56 & \text{when} & .2 < \dfrac{E_f}{E_c} \leq .5 \\[2mm] .71 & \text{when} & .5 < \dfrac{E_f}{E_c} \leq 1 \\[2mm] .83 & \text{when} & 1 < \dfrac{E_f}{E_c} \end{cases} \tag{33}$$

Fig. 12 Calculated behavior of hardening concrete sample and crack identification (Schindler et al. 2010)

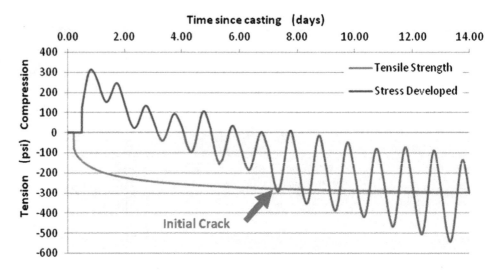

Table 5 HIPERPAV versus deck cracking spreadsheet comparison summary

Case study (no—description)	HIPERPAV result (for concrete roadway pavement)	Deck cracking spreadsheet result (for reinforced concrete deck)
I—Cement type	Type III cement: requires saw cut before 12 h	Type III cement: cracks at 6 days
	Type I cement + fly ash: requires saw cut before 15 h	Type I cement + fly ash: cracks at 15 days
II—Aggregate type	Low CTE value: has wider crack spacing	Low CTE value: cracks at later time
	High CTE value: has narrower crack spacing	High CTE value: cracks at earlier time

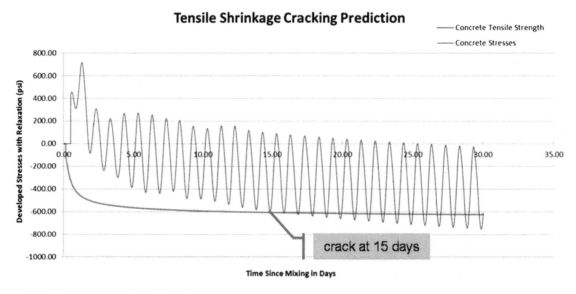

Fig. 13 Case study I: cement type I with fly ash

Total developed stresses

Using the stresses calculated from the strains in Eqs. 29, 30, and 31, the total stress in the newly casted deck can be calculated from Eq. 34.

$$\sigma_{\text{total}} = K_r \cdot (\sigma_J + \sigma_T + \sigma_{\text{cr}}) \tag{34}$$

Time of first developed crack

Finally, the tensile strength can be calculated by Eq. 35; where $f'c$ is as calculated in Eq. 26, and w is the calculated unit weight of the concrete determined from the mix design.

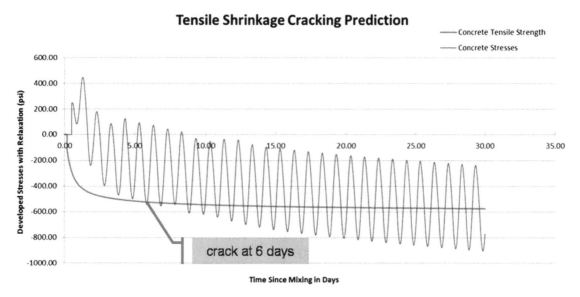

Fig. 14 Case study I: cement type III

$$f_t = \frac{\sqrt{f'c \cdot w}}{3} \qquad (35)$$

The age of the concrete at initial cracking of the deck can be approximated by comparing the developed tensile strength of the concrete to the stresses developed in the concrete. An example of this comparison can be seen in Fig. 10.

Similar to the reference documents, when relaxation effects are taken into consideration, a point of zero stress can be identified, and the moment in time of cracking is shown by the first intersection of the two graphed properties (developed strength and developed stresses). An example of the theory is depicted in Fig. 11 which was taken from Schindler (2002). This research developed a graph which follows the concept in Fig. 11, and it is shown in Fig. 12 where the age of concrete at time of first cracking can be easily identified.

Comparison to HIPERPAV

The HIPERPAV® (*HIgh PERformance Concrete PAVing*) software is used to analyze the early-age behavior of jointed concrete pavements, continuously reinforced concrete pavements, and bonded concrete overlays. It is important to compare results of developed spreadsheet to those of available tools, therefore, two HIPERPAV® case studies were run in the Deck Cracking Spreadsheet and the results were compared to results from HIPERPAV®. It was found that results have matched in most cases (Table 5).

Case study I

This HIPERPAV case study investigates how a change in cement type affects saw cutting and probability of cracking on a fast track project. Two concrete mix designs were analyzed, the first mix uses a Cement Type I with fly ash, and the second mix uses a Cement Type III. From results of HYPERPAV, it is anticipated that the mix using Type III Cement will develop cracks at an earlier age than the mix using Type I Cement with fly ash, and hence, the optimum time period for saw cutting may be reduced when Type III Cements are used. The results obtained from the Deck Cracking Spreadsheet, see Figs. 13 and 14, showed that concrete with Type I Cement and fly ash developed first crack at about 15 days, while concrete with Type III Cement developed first crack at about 6 days. It is clear that this finding agrees with performance observed when HYPERPAV was used.

Case study II

This HIPERPAV case study investigates how coarse aggregate type affects the performance of the CRCP for a set of climatic conditions. Five concrete mix designs were analyzed each with a different type of coarse aggregate. The Deck Cracking Spreadsheet is used to assess the behavior of an 11-inch CRCP constructed with concrete containing different aggregate types, namely siliceous river gravel (Lime), basalt, granite/gneiss, sandstone, and limestone. The climate is assumed to be for temperatures at noon on December 12. The contraction and expansion of the concrete deck depends on

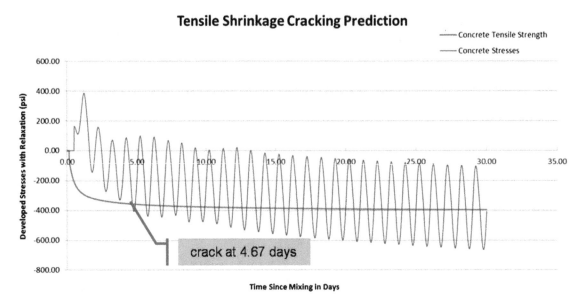

Fig. 15 Case study II: siliceous gravel

Table 6 Case study II: CTE results

Aggregate type	Limestone	Basalt	Granite	Sandstone	Siliceous gravel
CTE (με/°F)	2.6	3.7	4.2	6.2	6.5
Crack time (days)	23.76	4.69	8.75	4.69	4.67

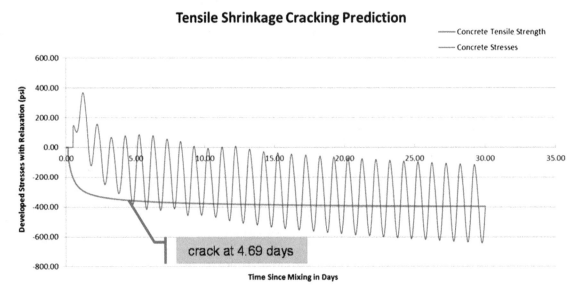

Fig. 16 Case study II: sandstone

the coarse aggregate coefficient of thermal expansion (CTE), since coarse aggregate comprises about half of the concrete volume. Since temperature changes are the greatest in the deck immediately after construction, its volume changes are significant at early ages. From the HIPERPAV results, it is anticipated that the pavement constructed with the low CTE aggregate provides better performance since it experiences lower thermal stresses. Results of spreadsheet for all the five cases (Fig. 15 through Fig. 19) are summarized in Table 6, and results show similar trend to those of HIPERPAV. With the exception of Basalt, which cracks at 4.7 days, all the other four types of coarse aggregates seem to follow the trend anticipated in the HIPERPAV models. The deck

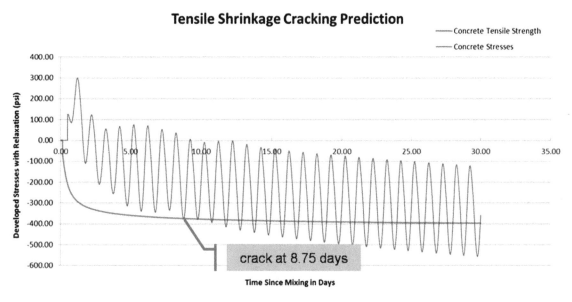

Fig. 17 Case study II: granite

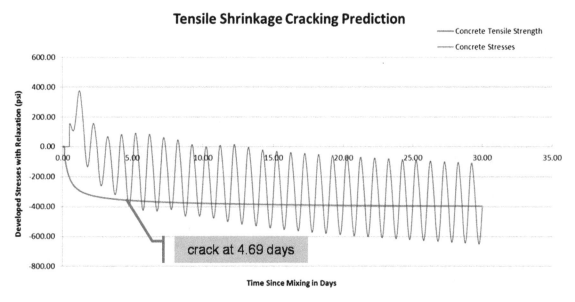

Fig. 18 Case study II: basalt

constructed with the low CTE aggregate provides better performance by cracking at a later time.

New deck analysis

The bridge under study consists of two simply-supported spans on AASHTO Type II beams. The deck is poured continuous over the intermediate bent.

The Deck and Concrete Input information is shown in Fig. 20. The project location is close to Lakeland Florida. Based on the design plans, the concrete deck is 8 inches

thick, the 28 day compressive strength is 4500 psi. The top and bottom reinforcing steel mats consist of #5 rebars spaced at eight inches.

The Structure and Environmental information and input are shown in Fig. 21. Based on the progress reports obtained from CEI, it appears that the deck pour took place in the morning of June 11, 2012. The weather conditions for the first 14 days after the deck pour were obtained. The information includes the maximum and minimum temperatures, the maximum wind speed, and the maximum and minimum relative humidity.

The result summary is shown in Fig. 22. The graph theoretically shows that no cracks are anticipated to take

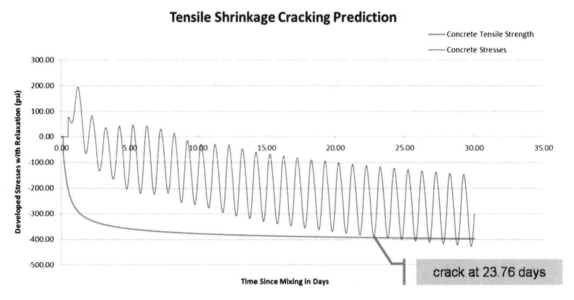

Fig. 19 Case study II: limestone

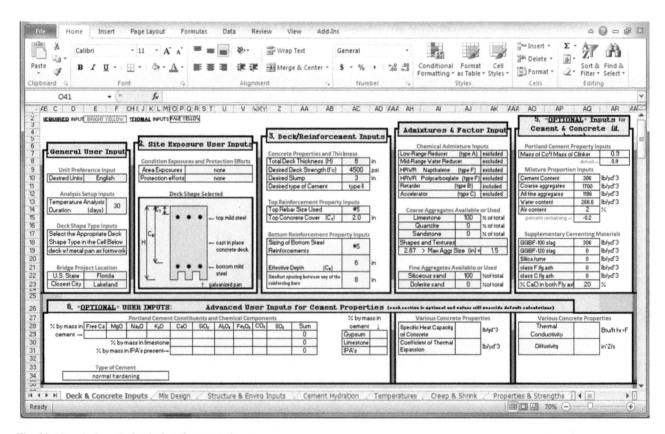

Fig. 20 New deck analysis: deck and concrete inputs

place in this new deck. This could be attributed to the following reasons:

1. The concrete mix design had taken into consideration hot weather concreting requirements.

2. The weather conditions seem to be favorable for deck pouring since the maximum relative humidity was over 95 % for 13 out of the 14 days after deck casting and the minimum relative humidity registered at 95 % on day 14.

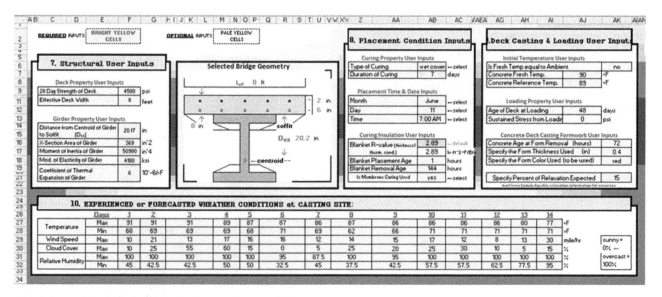

Fig. 21 New deck analysis: structure and environmental inputs

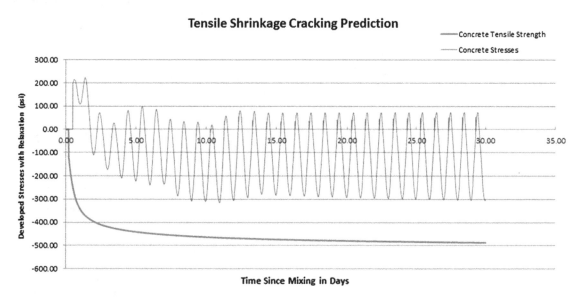

Fig. 22 New deck analysis: result summary

Conclusions

There are several conclusions drawn from this study including implementing a number of practical methods for evaluating and reducing the risk of early-age cracking such as reducing the placement temperature of the concrete, selecting an aggregate with a low coefficient of thermal expansion, using a favorable grading, using a large maximum size aggregate, using a relatively coarsely ground cement with a low alkali content, and a high sulfate content relative to its C_3A content, substituting some of the cement with fly ash, using entrained air, and using SRAs.

A tool was developed to predict transverse deck cracking based on the properties of the bridge deck, mix design,

and environment and the types of loads applied. The availability of such tool is expected to make the evaluation of likelihood of cracking in bridge decks more efficient, since it displays both the developed stresses and when cracks may take place.

The Deck Cracking Spreadsheet is used for concrete bridge decks and addresses a few different types of construction including decks with stay-in-place galvanized forms, decks with removable forms, and decks on precast panels. The program can be accessed through FDOT. The Deck Cracking Spreadsheet is a user-friendly calculation tool for concrete mixture proportioning, temperature prediction, thermal analysis, and tensile cracking prediction. The Deck Cracking Spreadsheet results were also compared with the HIPERPAV® software results by conducting a number of case studies.

Also, users can arrive at a crack-free design for bridge deck using the developed tool by inputting the concrete mix design, taking into consideration hot weather concreting requirements, and favorable weather conditions. The tool's graph theoretically can show that no cracks are anticipated to take place in this designed concrete bridge deck.

References

Al-Fadhala M, Hover KC (2001) Rapid evaporation from freshly cast concrete and the Gulf environment. Constr Build Mater 15(1):1–7

American Association of State Highway and Transportation Officials (AASHTO) (2007) LRFD bridge design specifications, 4th Edition, Washington DC

American Concrete Institute (2009) ACI Committee 211: "Standard practice for selecting proportions for normal, heavyweight and mass concrete". American Concrete Institute, Farmington Hills

American Concrete Institute (2010) ACI Committee 231, "report on early-age cracking: causes, measurement, and mitigation (ACI 231R–10)". American Concrete Institute, Farmington Hills, p 46

American Concrete Institute Committee 301, Specifications for Structural Concrete 2010

Babaei K, Hawkins NM (1987) Evaluation of bridge deck protective strategies. NCHRP report 297, TRB, National Research Council, Washington, DC

Babaei K, Purvis RL (1996) Premature cracking of concrete bridge decks: cause and methods of prevention. In: Proceedings, 4th international bridge engineering conference

Darwin D, Browning J, Mcleod HAK, Lindquist W, Yuan J (2012) Implementing lessons learned from twenty years of bridge-deck crack surveys. ACI Spec Publ 284:1–18

Deschutter G, Taerwe L (1996) Estimation of early-age thermal cracking tendency of massive concrete elements by means of equivalent thickness. ACI Mater J 93(5):403–408

El Safty A, Abdel-Mohti A, Jackson NM, Lasa I, Paredes M (2013) Limiting early-age cracking in concrete bridge decks. Adv Civil Eng Mater 2(1):379–399. doi:10.1520/ACEM20130073 **ISSN 2165-3984**

ElSafty A, Abdel-Mohti A (2013) Investigation of likelihood of cracking in reinforced concrete bridge decks. Int J Concr Struct Mater (IJCSM), 7(1):79–93

ElSafty A, Jackson NM (2012) Sealing of cracks on florida bridge decks with steel girders. Report, FDOT Contract #BDK82 977-02, FDOT, State Research Office, Tallahassee

Emanuel JH, Hulsey JL (1977) Prediction of the thermal coefficient of expansion of concrete. J Am Concr Inst 74(4):149–155

French C, Eppers L, Le Q, Hajjar JF (1999) Transverse cracking in concrete bridge decks. Transportation research record. n 1688, pp 21–29

Manafpour A, Hopper T, Radlinska A, Warn G, Rajabipour F, Morian D, Jahangirnejad S (2015) Bridge deck cracking: effects on in-service performance, prevention, and remediation. Pennsylvania Department of Transportation, Report

Hover KC (2006) Evaporation of water from concrete surfaces. ACI Mater J 103(5):384–389

Krauss PD, Rogalla EA (1996) Transverse cracking in newly constructed bridge decks. NCHRP Report 380, Transportation Research Board, Washington

La Fraugh RW, Perenchio WF (1989) Phase I report of bridge deck cracking study West Seattle Bridge. Report No 890716, Wiss, Janney, Elstner Associates, Northbrook, Ill

Maggenti R, Knapp C, Fereira S (2013) Controlling shrinkage cracking: available technologies can provide nearly crack-free concrete bridge decks. Concr Int 35(7):36–41

McLeod HAK, Darwin D, Browning J (2009) Development and construction of low-cracking high performance concrete (LC-HPC) bridge decks: construction methods, specifications, and resistance to chloride ion penetration. SM report 94. University of Kansas Center for Research, Lawrence

PCA (1970) Durability of concrete bridge decks. Final report, p 35

Peyton SW, Sanders CL, John EE, Hale WM (2012) Bridge deck cracking: a field study on concrete placement, curing, and performance. Constr Build Mater 34:70–76. doi:10.1016/j.conbuildmat.2012.02.065

Schindler AK (2002) Concrete hydration, temperature development, and setting at early-ages. Ph.D. Dissertation, The University of Texas at Austin, Texas

Schindler AK, Folliard KJ (2005) Heat of hydration models for cementitious materials. ACI Mater J 102(1):24–33

Schindler AK, Hughes ML, Barnes RW, Byard BE (2010) Evaluation of cracking of the US 331 bridge deck. Report No. FHWA/ALDOT 930-645, Highway Research Center, Auburn, AL, p 150

Schmitt TR, Darwin D (1999) Effect of material properties on cracking in bridge decks. J Bridge Eng ASCE 4(1):8–13

Slatnick S, Riding KA, Folliard KJ, Juenger MCG, Schindler A (2011) Evaluation of autogenous deformation of concrete at early ages. ACI Mater J 108(1):21–28

Van Breugel K (1998) Prediction of temperature development in hardening concrete. In: Springenschmid R, editor. Prevention of thermal cracking in concrete at early ages. E&FN Spon, London; pp 51–75. RILEM Report 15 [Chapter 4]

Wan B, Foley CM, Komp J (2010) Concrete cracking in new bridge decks and overlays. WHRP Report 10-05, Wisconsin Highway Research Program, p 154

Wright JR, Rajabipour F, Laman JA, Radlińska A (2014) Causes of early age cracking on concrete bridge deck expansion joint repair sections. Adv Civil Eng 2014:10. doi:10.1155/2014/103421

Florida Department of Transportation Standard Specifications for Road and Bridge Construction

Florida Department of Transportation Structures Manual Volume 1 Structures Design Guidelines

Ohio Department of Transportation Bridge Design Manual

Pennsylvania Department of Transportation Design Manual Part 4 Structures (DM-4)

6

Modified creep and shrinkage prediction model B3 for serviceability limit state analysis of composite slabs

Alireza Gholamhoseini[1]

Abstract Relatively little research has been reported on the time-dependent in-service behavior of composite concrete slabs with profiled steel decking as permanent formwork and little guidance is available for calculating long-term deflections. The drying shrinkage profile through the thickness of a composite slab is greatly affected by the impermeable steel deck at the slab soffit, and this has only recently been quantified. This paper presents the results of long-term laboratory tests on composite slabs subjected to both drying shrinkage and sustained loads. Based on laboratory measurements, a design model for the shrinkage strain profile through the thickness of a slab is proposed. The design model is based on some modifications to an existing creep and shrinkage prediction model B3. In addition, an analytical model is developed to calculate the time-dependent deflection of composite slabs taking into account the time-dependent effects of creep and shrinkage. The calculated deflections are shown to be in good agreement with the experimental measurements.

Keywords Composite slabs · Creep · Deflection · Profiled steel decking · Serviceability · Shrinkage

Introduction

Composite one-way concrete floor slabs with profiled steel decking as permanent formwork are commonly used in the construction of floors in buildings (Fig. 1a). The steel decking supports the wet concrete of a cast in situ reinforced or post-tensioned concrete slab and, after the concrete sets, acts as external reinforcement. Embossments on the profiled sheeting provide the necessary shear connection to ensure composite action between the concrete and the steel deck (Fig. 1b).

Despite their common usage, relatively little research has been reported on the in-service behavior of composite slabs. In particular, the drying shrinkage profile through the slab thickness (which is greatly affected by the impermeable steel decking) and the restraint to shrinkage provided by the steel decking have only recently been quantified (Gilbert et al. 2012; Ranzi et al. 2012; Al-deen and Ranzi 2015; Al-Deen et al. 2011; Ranzi et al. 2013). Carrier et al. (1975) measured the moisture contents of two bridge decks, one was a composite slab with profiled steel decking and the other was a conventional reinforced concrete slab permitted to dry from the top and bottom surfaces after the timber forms were removed. The moisture loss was significant only in the top 50 mm of the slab with profiled steel decking and in the top and bottom 50 mm of the conventionally reinforced slab. In their research, Gilbert et al. (2012) measured the nonlinear variation of shrinkage strain through the thickness of several slab specimens, with and without steel decking at the soffit, and sealed on all exposed concrete surfaces except for the top surface. Ranzi et al. (2012) carried out long-term tests on a post-tensioned solid concrete slab and two composite slabs with two different steel decking types and also measured the occurrence of non-uniform shrinkage strain through the thickness of the two composite slabs. Bradford (2010) presented a generic model for composite slabs subjected to concrete creep and two types of indirect (or non-mechanical) straining effects; shrinkage and thermal strains; including the effects of partial interaction between the concrete slab and steel decking.

✉ Alireza Gholamhoseini
alireza.gholamhoseini@gmail.com;
alireza.gholamhoseini@canterbury.ac.nz

[1] The University of Canterbury, Christchurch, New Zealand

(a) Soffit of a one-way composite slab and beam floor system

(b) Trapezoidal steel decking profile KF70

Fig. 1 Profiled steel decks (Fielders Australia)

As a consequence of the dearth of published research, little design guidance is available to structural engineers for predicting the in-service deformation of composite slabs. The techniques used to predict deflection and the on-set of cracking in conventionally reinforced concrete slabs (Gilbert 1999; Gilbert and Ranzi 2011) are often applied inappropriately. Although techniques are available for the time-dependent analysis of composite slabs (Gilbert and Ranzi 2011), due to lack of guidance in codes of practice, structural designers often specify the decking as sacrificial formwork, in lieu of timber formwork, and ignore the structural benefits afforded by the composite action. Of course this provides a conservative estimate of ultimate strength of the slab and is quite unsustainable, but may well result in a significant under-estimation of deflection because of the shrinkage strain gradient and the restraint provided by the deck and this should not be ignored.

In this paper, the results of an experimental study of the long-term deflection of composite concrete slabs due to sustained service loads and shrinkage are presented. Deflections caused by creep of the concrete and the effects of drying shrinkage are reported and discussed. Based on the experimental results, a shrinkage strain profile is proposed based on some modifications to Bažant-Baweja B3 model (ACI Committee 209 2008) for prediction of creep and shrinkage for design purposes and an analytical technique is proposed for determining the time-varying deflections of composite floor slabs with profiled steel decking. Good agreement is obtained between the calculated and measured deflections.

Bažant-Baweja B3 model

The Bažant-Baweja (1995) B3 model is the most recent model among a number of shrinkage and creep prediction methods developed over the years by Bažant et al. and is restricted to portland cement concrete cured for at least 1 day and to the service stress range of up to $0.45f_{cm28}$ with the following ranges of parameters:

$$0.35 \leq w/c \leq 0.85$$

(water-cement ratio by weight).

$$2.5 \leq a/c \leq 13.5$$

(aggregate-cement ratio by weight).

$$17 \text{ MPa} \leq f_{cm28} \leq 70 \text{ MPa}$$

(28-day mean standard cylinder concrete compressive strength).

$$160 \text{ kg/m}^3 \leq c \leq 720 \text{ kg/m}^3$$

(cement content)

If only the characteristic design strength of concrete is known, then $f_{cm28} = f_c' + 8.3$ MPa is taken as the default value.

Design shrinkage strain

In the Bažant-Baweja B3 model, the mean shrinkage strain $\varepsilon_{sh}(t, t_c)$ in the cross section at age of concrete t (in days) that has developed since the start of drying at age t_c (in days) is:

$$\varepsilon_{sh}(t, t_c) = \varepsilon_{shu}k_h S(t - t_c) \tag{1}$$

where ε_{shu} is the ultimate shrinkage strain; k_h is the humidity dependence factor; $S(t - t_c)$ is the time function for shrinkage strain; and $(t - t_c)$ is the drying period from the end of the initial curing. The ultimate shrinkage ε_{shu} is given by:

$$\varepsilon_{shu} = \varepsilon_{su}\frac{E_{cm(7+600)}}{E_{cm(t_c+\tau_{sh})}} \tag{2}$$

where ε_{su} is a constant strain given by Eq. (3), and $\frac{E_{cm(7+600)}}{E_{cm(t_c+\tau_{sh})}}$ is a factor that represents the time dependence of the ultimate shrinkage and is calculated using Eq. (4).

$$\varepsilon_{su} = \alpha_1\alpha_2\left[0.019w^{2.1}f_{cm28}^{-0.28} + 270\right] \times 10^{-6} \tag{3}$$

$$E_{cmt} = E_{cm28}\left(\frac{t}{4 + 0.85t}\right)^{0.5} \tag{4}$$

where w is the water content (in kg/m^3); f_{cm28} is the concrete mean compressive strength at 28 days (in MPa); and α_1 and α_2 are constant values associated with the cement type and curing condition, respectively. The value of α_1 is 1.0, 0.85 and 1.1 for cement types I, II and III, respectively.

The value of α_2 is 0.75, 1.0 and 1.2 for steam cured, cured in water or at 100 % relative humidity and for sealed during curing or normal curing in air with initial protection against drying, respectively.

The humidity dependence factor k_h is calculated from Eqs. (5) and (6) as:

$$k_h = 1 - \left(\frac{RH}{100}\right)^3 \quad \text{if } RH \leq 98\ \% \tag{5}$$

$$k_h = -0.2 \quad \text{if } RH = 100\ \% \tag{6}$$

and is obtained from linear interpolation between Eqs. (5) and (6) when 98 % < RH < 100 %.

The time function for shrinkage strain $S(t-t_c)$ is given by:

$$S(t-t_c) = \tanh \sqrt{\frac{(t-t_c)}{\tau_{sh}}} \tag{7}$$

$$\tau_{sh} = 0.085 t_c^{-0.08} f_{cm28}^{-0.25} [2k_s(V/S)]^2 \tag{8}$$

where V/S is the volume-surface ratio (in mm) and k_s is the shape-correction factor for the cross-section which is 1.0 for an infinite slab, 1.15 for an infinite cylinder, 1.25 for an infinite square prism, 1.3 for a sphere and 1.55 for a cube, respectively.

Compliance factor

Unlike the creep models in ACI 209R-92 (ACI Committee 209 2008) and in Australian Standard (AS3600-2009) (Australia 2009), the Bažant-Baweja B3 model does not predict the creep coefficient, but instead gives the average compliance factor $J(t,t_0)$ at concrete age t caused by a unit uniaxial constant stress applied at age t_0, which is the summation of the instantaneous strain, the basic creep strain and the drying creep strain:

$$J(t,t_0) = q_1 + C_0(t,t_0) + C_d(t,t_0,t_c) \tag{9}$$

where q_1 is the instantaneous strain due to a unit stress; $C_0(t,t_0)$ is the compliance factor related to basic creep; $C_d(t,t_0,t_c)$ is the additional compliance factor related to drying creep; and t, t_c, and t_0 are the age of concrete, the age that drying begins (or end of moist curing), and the age of concrete at the time of loading, respectively (all in days).

The instantaneous strain due to unit stress is expressed as:

$$q_1 = \frac{0.6}{E_{cm28}} \tag{10}$$

where

$$E_{cm28} = 4734\sqrt{f_{cm28}} \tag{11}$$

The compliance factor related to basic creep is calculated as:

$$C_0(t,t_0) = q_2 Q(t,t_0) + q_3 \ln[1 + (t-t_0)^n] + q_4 \ln\left(\frac{t}{t_0}\right) \tag{12}$$

where

$$q_2 = 185.4 \times 10^{-6} c^{0.5} f_{cm28}^{-0.9} \tag{13}$$

$$Q(t,t_0) = Q_f(t_0) \left[1 + \left(\frac{Q_f(t_0)}{Z(t,t_0)}\right)^{r(t_0)}\right]^{-1/r(t_0)} \tag{14}$$

$$Q_f(t_0) = \left[0.086(t_0)^{2/9} + 1.21(t_0)^{4/9}\right]^{-1} \tag{15}$$

$$Z(t,t_0) = (t_0)^{-m} \ln[1 + (t-t_0)^n] \tag{16}$$

$$r(t_0) = 1.7(t_0)^{0.12} + 8 \tag{17}$$

$$q_3 = 0.29(w/c)^4 q_2 \tag{18}$$

$$q_4 = 20.3(a/c)^{-0.7} \times 10^{-6} \tag{19}$$

where c is the cement content (in kg/m^3); f_{cm28} is the concrete mean compressive strength at 28 days (in MPa); m and n are empirical parameters with the values of $m = 0.5$ and $n = 0.1$ for all normal strength concretes; w/c is water-cement ratio; and a/c is aggregate-cement ratio, respectively.

The compliance factor related to drying creep is obtained from Eqs. (20)–(25) as:

$$C_d(t,t_0,t_c) = q_5 \left[e^{\{-8H(t)\}} - e^{\{-8H(t_0)\}}\right]^{0.5} \tag{20}$$

where

$$q_5 = 0.757 f_{cm28}^{-1} |\varepsilon_{shu} \times 10^6|^{-0.6} \tag{21}$$

$$H(t) = 1 - \left(1 - \left(\frac{RH}{100}\right)\right) S(t-t_c) \tag{22}$$

$$S(t-t_c) = \tanh \sqrt{\frac{(t-t_c)}{\tau_{sh}}} \tag{23}$$

$$H(t_0) = 1 - \left(1 - \left(\frac{RH}{100}\right)\right) S(t_0-t_c) \tag{24}$$

$$S(t_0-t_c) = \tanh \sqrt{\frac{(t_0-t_c)}{\tau_{sh}}} \tag{25}$$

And the creep coefficient can be calculated as:

$$\phi(t,t_0) = E_{cm28}(C_0(t,t_0) + C_d(t,t_0,t_c)) \tag{26}$$

Experimental program

Overview

The experimental program involved the testing of ten large scale simple-span composite one-way slabs under different sustained, uniformly distributed service load histories for periods of up to 244 days. Two different decking profiles KF40 and KF70 (Fielders Australia 2008) were considered as shown in Fig. 2.

The creep coefficient and drying shrinkage strain for the concrete were measured on companion specimens cast with the slabs and cured similarly. Additionally, the compressive strength and the elastic modulus of concrete at the age of first loading and at the end of the sustained load period were measured on standard 100 mm diameter cylinders; while the concrete flexural tensile strength (modulus of rupture) was measured on 100 mm × 100 mm × 500 mm concrete prisms. The elastic modulus E_{sd} and the yield stress f_y of the steel decking were also measured on coupons cut from the decking.

Crack locations and crack widths on the side surfaces of the slabs were recorded throughout the long-term test, together with the time-dependent change in concrete and steel strains, mid-span deflection and the slip between the steel decking and the concrete at each end of the specimen.

The objectives of the experimental program were to obtain benchmark, laboratory-controlled data on the long-term structural response of composite slabs under different sustained service loads, in particular the time-varying deflection, and to analyze the effect of creep and shrinkage on the long-term behavior of composite slabs. The laboratory data was then used to validate analytical models for the prediction of time-dependent behavior (Gilbert et al. 2012; Gilbert and Ranzi 2011) and to assist in the development of design-oriented procedures to assess the serviceability of composite slabs.

Test specimens and instrumentation

Each slab was 3300 mm long, with a cross-section 150 mm deep and 1200 mm wide, and contained no reinforcement (other than the external steel decking). Each slab was tested as a single simply-supported span. The center to center distance between the two end supports (one hinge and one roller) was 3100 mm. Five identical slabs with KF70 decking were poured at the same time from the same batch of concrete. An additional five identical slabs with KF40 decking were poured at a different time from a different batch of concrete (but to the same specification and from the same supplier). The thickness of the steel sheeting in both types of decking was $t_{sd} = 0.75$ mm. The cross-section of each of the five slabs with KF70 decking is shown in Fig. 3a. The choice of specimen variables was made in order to examine the effects of shrinkage and sustained load levels on long-term deflections for slabs with two different deck profiles, while keeping slab thickness and concrete properties the same for each specimen. Further testing will be necessary to consider the effects of varying the concrete properties and slab thickness on long-term deflection.

Each slab was covered with wet hessian and plastic sheets within 4 h of casting and kept moist for 6 days to delay the commencement of drying. At age 7 days, the side forms were removed and the slabs were lifted onto the supports. Subsequently, the slabs were subjected to different levels of sustained loading provided by means of different sized concrete blocks. A photograph of the five KF70 slabs showing the different loading arrangements and the slab designations are also shown in Fig. 4. The first digit in the designation of each slab is the specimen number (1–10) and the following two letters indicate the nature of the test, with LT for long-term. The next two numbers indicate the type of decking (with 70 and 40 for KF70 and KF40, respectively). The final digit indicates the approximate value of the maximum superimposed sustained loading in kPa.

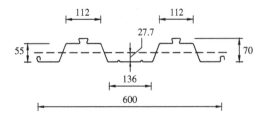

$A_{sd} = 1100$ mm^2/m; $y_{sd} = 27.7$ mm; $I_{sd} = 584000$ mm^4/m

(a) KF70

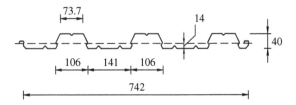

$A_{sd} = 1040$ mm^2/m; $y_{sd} = 14.0$ mm; $I_{sd} = 269000$ mm^4/m

(b) KF40

Fig. 2 Dimensions (in mm) of each steel decking profile ($t_{sd} = 0.75$ mm)

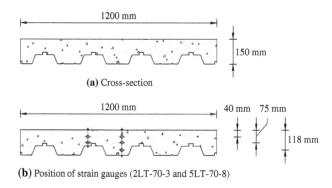

(a) Cross-section

1200 mm

150 mm

1200 mm   40 mm   75 mm

118 mm

(b) Position of strain gauges (2LT-70-3 and 5LT-70-8)

Fig. 3 Cross-sections and embedded strain gauges location in KF70 slabs

Fig. 4 View of slabs with KF70 decking under sustained load

The mid-span deflection of each slab was measured throughout the sustained load period with dial gauges at the soffit of the specimen. Dial gauges were also used to measure the slip between concrete slab and steel decking at the ends of the slab at both roller and hinge supports in slabs 2LT-70-3, 3LT-70-3, 4LT-70-6 and 5LT-70-8 with KF70 decking and in slabs 7LT-40-3 and 9LT-40-6 with KF40 decking. At the mid-span of each slab, the concrete strains were measured on the top and bottom surfaces using 60 mm long strain gauges. The strain gauges were glued onto the concrete surface and steel sheeting after removing the wet hessian at age 7 days. Internal embedded wire strain gauges were used to measure the concrete strains at different depths through the thickness of slabs 2LT-70-3, 5LT-70-8, 7LT-40-3 and 9LT-40-6, with locations shown on the cross-section in Fig. 3b. The self-weight and cross-sectional properties of the composite slabs are given in Table 1.

The location, height and width of the cracks were measured on the side faces of each specimen and recorded throughout the test. Of particular interest was the time-dependent development of cracking and the increase in crack widths with time. Crack widths were measured using a microscope with a magnification factor of 40. The average relative humidity *RH* in the laboratory throughout the period of testing was 67 and 72 % for the KF70 and KF40 test specimens, respectively.

Loading procedure

Each of the KF70 slabs was placed onto its supports at age 7 days and remained unloaded (except for its self-weight, see Table 1) until age 64 days. At age 64 days, with the exception of 1LT-70-0, each slab was subjected to super-imposed sustained loads in the form of concrete blocks. Each concrete block was placed on 60 mm high timber blocks to ensure a largely uninterrupted air flow over the top surface of the slabs and allow the concrete to shrink freely on the top surface. The block layouts are illustrated in Fig. 5 (and are also shown in the photograph of Fig. 4). Slab 1LT-70-0 carried only self-weight for the full test duration of 240 days. Slabs 2LT-70-3 and 3LT-70-3 were identical, carrying a constant superimposed sustained load of 3.4 kPa from age 64 to 247 days, i.e. a total sustained load of 6.4 kPa. Slab 4LT-70-6 carried a constant super-imposed sustained load of 6.0 kPa from age 64 to 247 days, i.e. a total sustained load of 9.0 kPa. Slab 5LT-70-8 carried a constant superimposed sustained load of 6.1 kPa from age 64 to 197 days, i.e. a total sustained load of 9.1 kPa and from age 197 to 247 days the superimposed sustained load was 7.9 kPa, i.e. a total sustained load of 10.9 kPa.

Each of the KF40 slabs was placed onto the supports at age 7 days and remained unloaded except for its self-weight, i.e. 3.2 kPa until age 28 days. At age 28 days (after 21 days drying), with the exception of 6LT-40-0, each slab was subjected to superimposed sustained loads with the block layouts similar to that used for the KF70 slabs and shown in Fig. 5. Slab 6LT-40-0 carried only self-weight for the full test duration of 244 days. Slabs 7LT-40-3 and 8LT-40-3 were identical, carrying a constant superimposed sustained load of 3.4 kPa from age 28 to 251 days, i.e. a total sustained load of 6.6 kPa. Slabs 9LT-40-6 and 10LT-40-6 were also identical and carried a constant superim-posed sustained load of 6.4 kPa from age 28 to 251 days, i.e. a total sustained load of 9.6 kPa.

Experimental results

Material properties

The measured compressive strength, modulus of elasticity and flexural tensile strength are presented in Table 2. The

Table 1 Properties of composite slabs

Slab decking profile	Slab self-weight (kPa)	Gross section I_g (mm^4)	Cracked section I_{cr} (mm^4)
KF70	3.0	278×10^6	102×10^6
KF40	3.2	310×10^6	111×10^6

measured creep coefficient versus time curves for concrete cylinders cast with the KF70 slabs and first loaded at age 64 days and KF40 slabs first loaded at age 28 days is shown in Fig. 6. The creep coefficient at the end of test for the KF70 slabs was ϕ (247,64) = 1.62. For the KF40 slabs, the creep coefficient at the end of the test (age 251 days) for the concrete first loaded at age 28 days was ϕ (251,28) = 1.50.

The development of the drying shrinkage strain for the concrete is also shown in Fig. 6. The curves represent the average of the measured shrinkage on two standard shrinkage prisms, 75 mm × 75 mm × 275 mm, from the day after removing the wet hessian until the end of the test. The average measured shrinkage strain at the end of test for the KF70 slabs was ε_{sh} = 512 µε. Similarly, for the KF40 slabs, the average measured shrinkage strain at the end of tests was ε_{sh} = 630 µε.

The average of the measured values of yield stress and elastic modulus taken from three test samples of the KF70 decking were f_y = 544 MPa and E_{sd} = 212 GPa, respectively. Similarly, from three test samples of the KF40 decking, average values were f_y = 475 MPa and E_{sd} = 193 GPa, respectively.

Mid-span deflection and end slip

The variations of mid-span deflection with time for the KF70 and KF40 slabs are shown in Fig. 7. Key deflection values are summarized in Table 3. The measured deflection includes that caused by shrinkage, the creep-induced deflection due to the sustained load (including self-weight), the short-term deflection caused by the superimposed loads (blocks) and the deflection caused by the loss of stiffness resulting from time-dependent cracking (if any). It does not include the initial deflection of the uncracked slab at age 7 days due to self-weight (which has been calculated to be about 0.5 mm for both the KF70 and KF40 slabs).

The measured end slips were very small with the maximum values of about 0.1 and 0.12 mm at the supports in 3LT-70-3 and 4LT-70-6, respectively. The end slips were negligible in the other slabs.

Discussion of test results

Shrinkage clearly has a dominant effect on the final deflection of these composite slabs. With a sustained load of 3.2 kPa (self-weight), the final deflection of 6LT-40-0

was 4.99 mm. When the sustained load was increased by a factor of about 3–9.6 kPa, the slabs suffered additional cracking and yet the final deflection only increased by a factor of about 1.4–6.94 mm (9LT-40-6) and by a factor of about 1.7–8.26 mm (10T-40-6). A similarly dominant effect of shrinkage over load was observed in the KF70 slabs.

Prior to the application of any load other than self-weight, the slabs deflected significantly, mainly due to the shrinkage-induced curvature. For the five KF70 slabs, after 57 days of drying (when ε_{sh} = 400 µε), the deflection varied from 2.18 mm (for 4LT-70-6) to 3.54 mm (for 2LT-70-3). Although this was mainly due to early shrinkage, it included the creep deflection resulting from self-weight which was estimated at about 0.4 mm. At this stage all KF70 slabs were identical (in terms of materials, geometry and load history), yet the deflection varied significantly. This highlights the large degree of variability when considering the service load behavior of concrete slabs, with deflection being highly dependent on the nonlinear and time-dependent behavior of the concrete. For the five KF40 slabs, after 21 days of drying (when ε_{sh} = 390 µε), the deflection varied from 2.72 mm (for 8LT-40-3) to 3.33 mm (for 7LT-40-3).

The extent of time-dependent cracking in the KF40 slabs was greater than that in the KF70 slabs and was somewhat unexpected. With the centroid of the KF40 steel decking being only 14 mm above the bottom of the slab (and that of the KF70 decking being 27.7 mm above the bottom), the tensile force that developed with time on the concrete, due to the restraint provided by the KF40 decking to drying shrinkage, is significantly more eccentric to the centroid of the concrete than that provided by the KF70 decking. This will increase the concrete tensile stress in the bottom fibers of the concrete and may have contributed to the observed differences in crack patterns.

Analytical modeling

An analytical procedure for the time-dependent analysis of composite concrete cross-sections with uniform shrinkage through the thickness of the concrete slab and with full interaction was presented by Gilbert and Ranzi (2011) using the age-adjusted effective modulus method (Dilger and Neville 1971; Bažant 1972). Gilbert et al. (2012) extended the method to calculate the effects of a non-

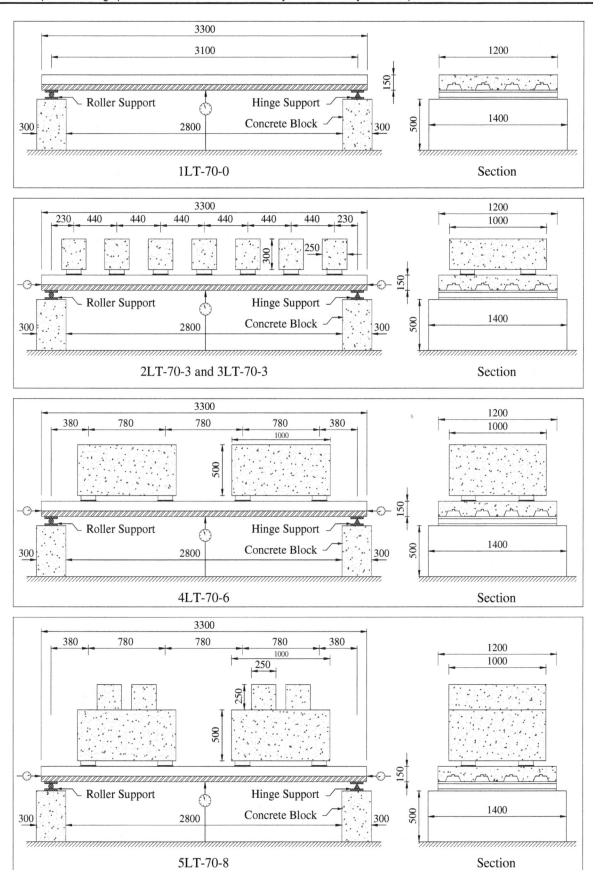

Fig. 5 Sustained load configuration for KF70 slabs

uniform shrinkage gradient by layering the concrete cross-section, with the shrinkage strain specified in each concrete layer depending on its position within the cross-section and with the assumption of full shear interaction at service load levels. This method is adopted here.

To calculate the time-dependent deformation of a composite concrete cross-section, the shrinkage strain profile and the creep coefficient for the concrete slab are needed. In the following, a shrinkage strain profile is proposed for concrete slabs on profiled steel decking that is suitable for use in structural design and a modification to the provisions of Bažant-Baweja B3 prediction model (ACI Committee 209 2008) for estimating the shrinkage strain and creep coefficient for composite slabs is also proposed. The proposals have been developed empirically from experimental measurements of shrinkage-induced strain distributions in composite slabs.

For a composite slab on profiled steel decking, if the average thickness of the concrete t_{ave} is defined as the area of the concrete part of the cross-section A_c divided by the width of the cross-section b, the following modifications to the term V/S is proposed to account for the effect of the steel decking on the drying profile through the concrete, and hence on the magnitude of creep and shrinkage:

$$V/S = 25 + 0.25t_{ave} \text{ (in mm)}. \tag{27}$$

Table 2 Concrete properties

Slab type	f_c' (MPa)	(days)	E_c (MPa)	(days)	$f_{ct.f}$ (MPa)	(days)
KF70	64	28.0	64	30,725	64	3.50
	247	29.8	247	31,650	247	4.54
KF40	28	35.5	28	28,200	28	3.80
	251	42.7	251	31,600	251	5.05

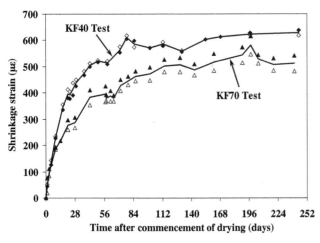
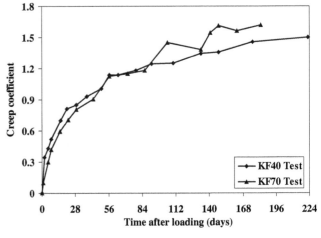

Fig. 6 Creep coefficient and shrinkage strain versus time

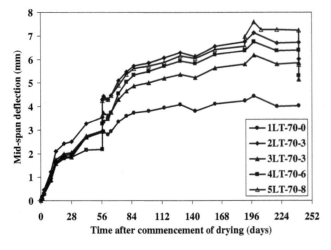

Fig. 7 Mid-span deflection versus time

Table 3 Measured mid-span deflections

Slab	Time-dependent deflection (mm)					
	57 days of drying		190 days of drying		240 days of drying	
	Before	After	Before	After	Before	After
1LT-70-0	2.92	2.92	4.24	4.24	4.04	4.04
2LT-70-3	3.54	4.29	6.74	6.74	6.72	6.01
3LT-70-3	2.97	3.63	5.80	5.80	5.84	5.16
4LT-70-6	2.18	3.38	6.37	6.37	6.40	5.31
5LT-70-8	2.94	4.23	6.56	6.96	7.23	5.78

Slab	Time-dependent deflection (mm)					
	21 days of drying		28 days of drying	56 days of drying	244 days of drying	
	Before	After			Before	After
6LT-40-0	2.83	2.83	3.15	3.87	4.99	4.99
7LT-40-3	3.33	4.14	4.72	5.68	7.30	6.62
8LT-40-3	2.72	4.12	4.70	5.38	6.57	5.53
9LT-40-6	2.95	4.35	4.60	5.90	6.94	5.68
10LT-40-6	3.30	5.10	5.52	6.72	8.26	7.81

For the decking profiles considered in this study, the ratio of trough height to slab thickness is defined as r_d (as described in Fig. 8) and was in the range 0.25–0.5. The modification factor K_m for creep coefficient and shrinkage strain is proposed as:

$$K_m = 1.5 - 0.55r_d \tag{28}$$

$$\varepsilon_{sh}^*(t, t_c) = K_m \varepsilon_{sh}(t, t_c) \tag{29}$$

$$\phi^*(t, t_0) = K_m \phi(t, t_0). \tag{30}$$

The measured shrinkage strain at any height y above the soffit of the composite slab with overall depth D, $\varepsilon_{sh}(y)$, may be approximated by Eq. (31):

$$\frac{\varepsilon_{sh}(y)}{\varepsilon_{sh}^*(t, t_c)} = \alpha + \beta\left(\frac{y}{D}\right)^4 \tag{31}$$

where $\varepsilon_{sh}(0) = \alpha \, \varepsilon_{sh}^*(t, t_c)$ is the shrinkage strain at the bottom of the slab (at $y = 0$) and $\varepsilon_{sh}(D) = (\alpha + \beta) \, \varepsilon_{sh}^*(t, t_c)$ is the shrinkage strain at the top surface of the slab (at $y = D$).

From the experimental results, $\alpha = 0.2$ provides a reasonable estimate, but β appears to depend on the profile of the steel decking. Excellent agreement between the predicted long-term deflection and the measured values is obtained with the value of $\beta = 2.0 - 2.25r_d$ as shown in Figs. 9 and 10.

The analytical curves were obtained by double integration of the curvature diagram at each time instant, with the curvature determined at cross-sections at 155 mm centers along the span using the layered cross-

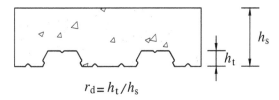

$$r_d = h_t / h_s$$

Fig. 8 Definition of shape factor (r_d)

section approach of Gilbert et al. (2012). Sample calculations for the determination of the short-term and time-dependent curvature at mid-span of slab 1LT-70-0 are provided in the Appendix, together with the determination of the shrinkage profile through the thickness of the slab.

For each slab, the same load history was considered in the analytical modeling as was applied to the real slab. Where two identical slabs with identical loading histories were tested, the analytical deflection-time curves are compared with the average of the two experimental curves. In those parts of the slabs where the numerical study showed that cracking had occurred, the effect of tension stiffening was considered using an approach similar to that outlined in Eurocode 2 (BS 1992). The average curvature (κ_{ave}) used in deflection calculation is determined according to Eq. (32):

$$\kappa_{ave} = \zeta \kappa_{cr} + (1 - \zeta)\kappa_{uncr} \tag{32}$$

where κ_{cr} is the time-dependent curvature on the cracked cross-section (ignoring tension in the concrete); κ_{uncr} is the

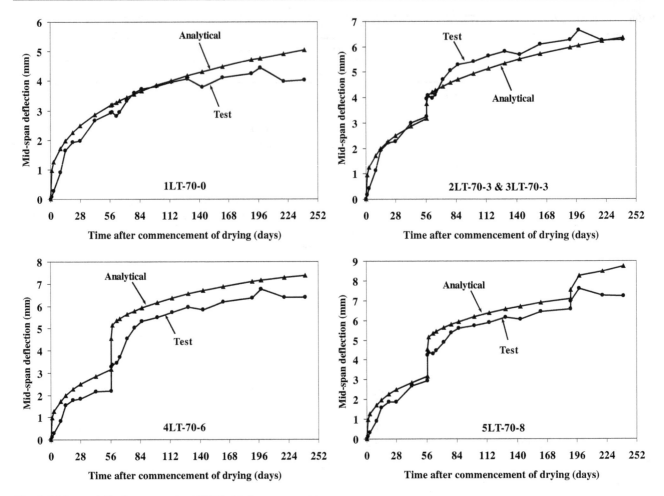

Fig. 9 Mid-span deflection versus time (KF70 slabs)

time-dependent curvature on the uncracked cross-section; and ζ is the distribution coefficient given by:

$$\zeta = 1 - \left(\frac{M_{cr.t}}{M_s}\right)^2 \tag{33}$$

where $M_{cr.t}$ is the cracking moment at the time under consideration and M_s is the in-service moment imposed on the cross-section.

The shrinkage induced deflection calculated using the proposed shrinkage profile for each decking type (Eq. 31), together with the instantaneous and time-dependent deflection caused by the applied load (elastic and creep deflection), are in good agreement with the measured response of the slabs.

A summary of the measured and calculated mid-span deflections is presented in Table 4, where comparisons between the measured and predicted deflections are made at 42 days after the commencement of drying and at the end of the test.

Summary and conclusions

The results of an experimental study of the long-term deflection of composite concrete slabs under sustained loads have been presented. The deformation caused by applied load, creep of the concrete and the effects of drying shrinkage have been reported and discussed for ten simply-supported slabs, with either KF70 or KF40 steel decking (Fielders Australia 2008), subjected to different loading histories. The measured slab deflections have confirmed the dominant effect of drying shrinkage over load for normal levels of sustained loads.

Also proposed is a nonlinear shrinkage profile through the thickness of a composite concrete slab, together with an analytical model for calculating the instantaneous and time-dependent curvature of the cross-section due to the effects of both load and nonlinear shrinkage. The agreement between the calculated deflection and the measured deflection for each of the ten slabs is good.

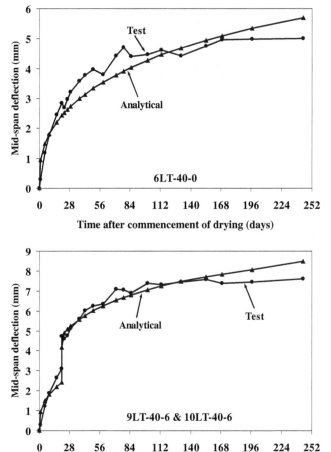

Fig. 10 Mid-span deflection versus time (KF40 slabs)

Appendix: Sample calculations

Sample calculations of the short-term and long-term curvature on the cross-section of slab 1LT-70-0 at mid-span are presented here using the approach presented by Gilbert et al. (2012). The slab was simply-supported over a span of 3100 mm and loaded with its self-weight of 3.60 kN/m at age 7 days. That uniform load remained constant for a further period of 240 days, with deformation increasing with time due to creep and shrinkage. The cross-section of the slab is divided into 10 layers, each 15 mm thick as shown in Fig. 11. Details of the geometric discretization are presented in Table 5. For this slab the elastic modulus

of the concrete is assumed to be constant as $E_c = 30.73$ GPa and for the steel decking $E_{sd} = 212$ GPa.

For this lightly loaded slab, the maximum sustained bending moment at mid-span is $M_{max} = 4.32$ kNm. The average thickness of the concrete is $t_{ave} = A_c/b = 148.8 \times 10^3/1200 = 124$ mm.

Take $t_c = t_0 = 7$ days, $t = 247$ days, $f_c = 28$ MPa, $f_{cm28} = f_c + 8.3 = 36.3$ MPa, $E_{cm28} = 30.73$ GPa, $\alpha_1 = 1.0$, $\alpha_2 = 1.2$, $c = 400$ kg/m^3, $w = 200$ kg/m^3, $a = 1650$ kg/m^3, $RH = 67$ %, $k_s = 1.0$.

Calculation of shrinkage strain in each concrete layer

From Eq. (4), $E_{cm607} = E_{cm28}\left(\frac{607}{4+0.85\times607}\right)^{0.5} = 33.20$ Gpa and with $t_c = 7$ days and $V/S = 25 + 0.25t_{ave} = 56$ mm, from Eq. (8) $\tau_{sh} = 0.085 \times 7^{-0.08} \times 36.3^{-0.25}[2 \times 1 \times 56]^2 = 371.8$ is calculated.

Table 4 Measured and calculated mid-span deflections

Slab	Time-dependent deflection (mm)					
	42 days of drying			240 days of drying		
	Measured (test)	Calculated	Measured/calculated	Measured (test)	Calculated	Measured/calculated
1LT-70-0	2.67	2.86	0.93	4.04	5.05	0.80
2LT-70-3	3.27	2.86	1.14	6.72	6.36	1.06
3LT-70-3	2.74	2.86	0.96	5.84	6.36	0.92
4LT-70-6	2.16	2.86	0.76	6.40	7.38	0.87
5LT-70-8	2.69	2.86	0.94	7.23	8.74	0.83

Slab	Time-dependent deflection (mm)					
	42 days of drying			244 days of drying		
	Measured (test)	Calculated	Measured/calculated	Measured (test)	calculated	Measured/calculated
6LT-40-0	3.77	3.13	1.20	4.99	5.70	0.88
7LT-40-3	5.50	4.16	1.32	7.30	7.04	1.04
8LT-40-3	5.26	4.16	1.26	6.57	7.04	0.93
9LT-40-6	5.67	5.76	0.98	6.94	8.47	0.82
10LT-40-6	6.36	5.76	1.10	8.26	8.47	0.98

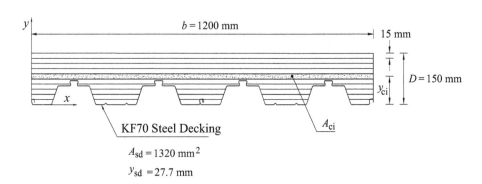

Fig. 11 Geometry and analysed cross-section of slab 1LT-70-0

$$b = 1200 \text{ mm}$$
$$15 \text{ mm}$$
$$D = 150 \text{ mm}$$
$$y_{ci}$$

KF70 Steel Decking

A_{ci}

$$A_{sd} = 1320 \text{ mm}^2$$
$$y_{sd} = 27.7 \text{ mm}$$

Table 5 Area, position and shrinkage strain of concrete layers

Layer (i)	$A_{c(i)}$ (mm^2)	y_{ci} (mm)	ε_{sh}
1	8768	7.5	-104.0×10^{-6}
2	9710	22.5	-104.3×10^{-6}
3	10,446	37.5	-105.9×10^{-6}
4	12,917	52.5	-111.4×10^{-6}
5	16,922	67.5	-124.3×10^{-6}
6	18,000	82.5	-149.2×10^{-6}
7	18,000	97.5	-192.2×10^{-6}
8	18,000	112.5	-260.3×10^{-6}
9	18,000	127.5	-361.9×10^{-6}
10	18,000	142.5	-506.4×10^{-6}
Σ	148,763		

$$E_{cm(t_c + \tau_{sh})} = E_{cm28} \left(\frac{378.8}{4 + 0.85 \times 378.8} \right)^{0.5} = 33.13 \text{ Gpa}$$

$$\varepsilon_{su} = 1 \times 1.2 \times [0.019 \times 200^{2.1} \times 36.3^{-0.28} + 270] \times 10^{-6}$$
$$= 891 \times 10^{-6}$$

$$S(t - t_c) = \tanh \sqrt{\frac{(247 - 7)}{371.7}} = 0.67$$

$$k_h = 1 - \left(\frac{67}{100} \right)^3 = 0.7$$

$$\varepsilon_{shu} = 891 \times 10^{-6} \times \frac{33.20}{33.13} = 893 \times 10^{-6}$$

$$\varepsilon_{sh}(t, t_c) = 893 \times 10^{-6} \times 0.7 \times 0.67 = 419 \times 10^{-6}$$

$$K_m = 1.5 - 0.55 r_d = 1.5 - 0.55 \times \frac{70}{150} = 1.24$$

$$\varepsilon_{sh}^*(t, t_c) = K_m \varepsilon_{sh}(t, t_c) = 1.24 \times 419 \times 10^{-6}$$
$$= 520 \times 10^{-6}$$

$$\beta = 2.0 - 2.25 r_d = 2 - 2.25 \times \frac{70}{150} = 0.95$$

The shrinkage strain at the centroid of the ith concrete layer on the cross-section is obtained from Eq. (31):

$$\varepsilon_{sh}(y_{ci}) = \varepsilon_{sh}^*(t, t_c) \times \left(\alpha + \beta \left(\frac{y_{ci}}{D}\right)^4\right)$$
$$= -520 \times \left(0.2 + 0.95 \left(\frac{y_{ci}}{150}\right)^4\right)$$

and is listed in Table 5.

Calculation of creep coefficient

From Eqs. (9) to (25), it can be found that:

$$q_2 = 185.4 \times 10^{-6} \times 400^{0.5} \times 36.3^{-0.9} = 146.3 \times 10^{-6}$$

$$Q_f(t_0) = \left[0.086 \times 7^{2/9} + 1.21 \times 7^{4/9}\right]^{-1} = 0.33$$

$$Z(t, t_0) = 7^{-0.5} \times \ln\left[1 + (247 - 7)^{0.1}\right] = 0.38$$

$$r(t_0) = 1.7 \times 7^{0.12} + 8 = 10.15$$

$$q_3 = 0.29 \times (200/400)^4 \times 146.3 \times 10^{-6} = 2.7 \times 10^{-6}$$

$$q_4 = 20.3 \times (1650/400)^{-0.7} \times 10^{-6} = 7.5 \times 10^{-6}$$

$$Q(t, t_0) = 0.33 \times \left[1 + \left(\frac{0.33}{0.38}\right)^{10.15}\right]^{-1/10.15} = 0.32$$

$$C_0(t, t_0) = 146.3 \times 10^{-6} \times 0.32 + 2.7 \times 10^{-6} \times \ln[1 + (247 - 7)^{0.1}] + 7.5 \times 10^{-6} \times \ln\left(\frac{247}{7}\right)$$
$$= 76.3 \times 10^{-6}$$

$$S(t_0 - t_c) = 0$$

$$H(t_0) = 1$$

$$H(t) = 1 - (1 - 0.67) \times 0.67 = 0.78$$

$$q_5 = 0.757 \times 36.3^{-1} \times 893^{-0.6} = 354 \times 10^{-6}$$

$$C_d(t, t_0, t_c) = 354 \times 10^{-6} \times [e^{-8 \times 0.78} - e^{-8}]^{0.5}$$
$$= 14.2 \times 10^{-6}$$

And the creep coefficient at age 247 days due to loading first applied at $t_0 = 7$ days is:

$$\phi(t, t_0) = 30.73 \times 10^3 \times \left(76.3 \times 10^{-6} + 14.2 \times 10^{-6}\right)$$
$$= 2.77$$

$$\overset{*}{\phi}(t, t_0) = 1.24 \times 2.77 = 3.43$$

Short-term analysis at mid-span at $t_0 = 7$ days

The strain at any point on the cross-section y above the slab soffit at time $t_0 = 7$ days, immediately after first loading can be expressed as $\varepsilon = \varepsilon_{r,0} - y\kappa_0$, where $\varepsilon_{r,0}$ is the strain at the slab soffit and κ_0 is the instantaneous curvature. Following the approach outline by Gilbert et al. (2012), the strain at mid-span due to any combination of axial force N_0 and moment M_0 is:

$$\begin{bmatrix} \varepsilon_{r,0} \\ \kappa_0 \end{bmatrix} = \frac{1}{R_{A,0}R_{I,0} - R_{B,0}^2} \begin{bmatrix} R_{I,0} & R_{B,0} \\ R_{B,0} & R_{A,0} \end{bmatrix} \times \begin{bmatrix} N_0 \\ M_0 \end{bmatrix} \quad (34)$$

where $R_{A,0}$, $R_{B,0}$ and $R_{I,0}$ are the rigidities of the transformed section related to area, first moment of area and second moment of area about the slab soffit and are calculated as:

$$R_{A,0} = \sum_{i=1}^{10} A_{ci}E_c + E_{sd}A_{sd} = 4851 \times 10^6 \text{ N}$$

$$R_{B,0} = \sum_{i=1}^{10} y_{ci}A_{ci}E_c + y_{sd}A_{sd}E_{sd} = 396 \times 10^9 \text{ Nmm}$$

$$R_{I,0} = \sum_{i=1}^{10} y_{ci}^2 A_{ci}E_c + (y_{sd}^2 A_{sd} + I_{sd})E_{sd}$$
$$= 40.7 \times 10^{12} \text{ Nmm}^2$$

when $N_0 = 0$ and $M_0 = 4.32$ kNm, Eq. (34) gives:

$\varepsilon_{r,0} = 42.1 \times 10^{-6}$ and $\kappa_0 = 0.52 \times 10^{-6} \text{mm}^{-1}$.

The strain in the bottom concrete layer at $y = 0$ mm is $\varepsilon = \varepsilon_{r,0} - y\kappa_0 = 42.1 \times 10^{-6}$ and the corresponding concrete stress is $\sigma_c = \varepsilon E_c = 1.3$ MPa. Since this is well below the tensile strength of concrete, this slab has not cracked at this time.

Long-term analysis at mid-span at $t_k = 247$ days

Adopting the age-adjusted effective modulus method as outlined by Gilbert et al. (2012), with an aging coefficient of $\chi(t_k, t_0) = 0.65$, the age-adjusted modulus for concrete after 240 days under load is:

$$\bar{E}_{e,k} = \frac{E_c}{1 + \chi(t_k, t_0)\phi(t_k, t_0)} = \frac{30725}{1 + 0.65 \times 3.43}$$
$$= 9514 \text{ MPa}.$$

The strain at any point on the cross-section y above the slab soffit at time $t_k = 247$ days can be expressed as $\varepsilon = \varepsilon_{r,k} - y\kappa_k$, where $\varepsilon_{r,k}$ is the strain at the slab soffit and κ_k is the curvature at time t_k. Following the approach outline by Gilbert et al. (2012), the strain at mid-span due to any combination of sustained axial force N_k and moment M_k is:

$$\begin{bmatrix} \varepsilon_{r,k} \\ \kappa_k \end{bmatrix} = \frac{1}{R_{A,k}R_{I,k} - R_{B,k}^2} \begin{bmatrix} R_{I,k} & R_{B,k} \\ R_{B,k} & R_{A,k} \end{bmatrix} \times (\mathbf{r}_k - \mathbf{f}_{cr,k} + \mathbf{f}_{sh,k})$$

(35)

where $R_{A,k}$, $R_{B,k}$ and $R_{I,k}$ are the rigidities of the age-adjusted transformed section related to area, first moment of area and second moment of area about the slab soffit and are calculated as:

$$R_{A,k} = \sum_{i=1}^{10} A_{ci}\bar{E}_{e,k} + E_{sd}A_{sd} = 1695 \times 10^6 \text{ N}$$

$$R_{B,k} = \sum_{i=1}^{10} y_{ci}A_{ci}\bar{E}_{e,k} + y_{sd}A_{sd}E_{sd} = 128 \times 10^9 \text{ Nmm}$$

$$R_{I,k} = \sum_{i=1}^{10} y_{ci}^2 A_{ci}\bar{E}_{e,k} + (y_{sd}^2 A_{sd} + I_{sd})E_{sd}$$
$$= 12.8 \times 10^{12} \text{ Nmm}^2$$

The vectors $\mathbf{r}_k$, $\mathbf{f}_{cr,k}$ and $\mathbf{f}_{sh,k}$ are vectors of axial force and moment: with $\mathbf{r}_k$ consisting of N_k and M_k; $\mathbf{f}_{cr,k}$ contains the fictitious actions resulting from the change in strain caused by creep due to the initial concrete stress at age t_0 and assuming full restraint; and $\mathbf{f}_{sh,k}$ contains the actions if the shrinkage strain was completely restrained over the time period.

$$\mathbf{f}_{cr,k} = \sum_{i=1}^{m_c} \bar{F}_{e(i),0} \begin{bmatrix} N_{c(i),0} \\ M_{c(i),0} \end{bmatrix}$$
$$= \sum_{i=1}^{m_c} \bar{F}_{e(i),0} E_{c(i),0} \begin{bmatrix} A_{c(i)}\varepsilon_{r,0} - B_{c(i)}\kappa_0 \\ -B_{c(i)}\varepsilon_{r,0} + I_{c(i)}\kappa_0 \end{bmatrix}$$

(36)

$$\mathbf{f}_{sh,k} = \sum_{i=1}^{m_c} \begin{bmatrix} A_{c(i)} \\ -B_{c(i)} \end{bmatrix} \bar{E}_{e(i),k}\varepsilon_{sh(i),k}$$

(37)

where the terms $A_{c(i)}$, $B_{c(i)}$ and $I_{c(i)}$ are the area, the first and second moments of area of the ith concrete layer about the x axis, respectively. In this case:

$$\mathbf{r}_k = \begin{bmatrix} 0 \\ 4.32 \times 10^6 \text{ Nmm} \end{bmatrix};$$

$$\mathbf{f}_{cr,k} = \begin{bmatrix} 3412 \text{ N} \\ -1.72 \times 10^6 \text{ Nmm} \end{bmatrix};$$

$$\mathbf{f}_{sh,k} = \begin{bmatrix} -314 \times 10^3 \text{ N} \\ 33.3 \times 10^6 \text{ Nmm} \end{bmatrix}$$

and from Eq. (35):

$$\varepsilon_{r,k} = 183 \times 10^{-6} \quad \text{and} \quad \kappa_k = 4.90 \times 10^{-6} \text{ mm}^{-1}.$$

Similar calculations may be formed at other cross-sections along the member and the curvature diagrams at times t_0 and t_k may be integrated to determine the slab deflection at each time, namely Δ_0 and Δ_k.

Table 6 Calculated short-term and long-term curvatures

Section (i)	x_i (mm)	$\kappa_0 \times 10^{-6}$ (mm^{-1})	$\kappa_k \times 10^{-6}$ (mm^{-1})
0	0	0.00	3.10
1	155	0.10	3.44
2	310	0.18	3.75
3	465	0.26	4.02
4	620	0.33	4.25
5	775	0.38	4.45
6	930	0.43	4.61
7	1085	0.47	4.74
8	1240	0.49	4.83
9	1395	0.51	4.89
10	1550	0.52	4.90

The curvatures so determined at ten points along the span are given in Table 6, where x_i is the distance of the section from the left end support of the slab.

Integration of the curvatures at each time instant gives the mid-span deflection:

$$\Delta_0 = 0.51 \text{ mm and } \Delta_k = 5.52 \text{ mm}.$$

The time-dependent part of the mid-span deflection is therefore: $\Delta_k^* = 5.52 - 0.51 = 5.01$ mm.

References

ACI Committee 209 (2008) Guide for modeling and calculating shrinkage and creep in hardened concrete (ACI 209.2R-08). American Concrete Institute, Farmington Hills

Al-deen S, Ranzi G (2015) Effects of non-uniform shrinkage on the long-term behaviour of composite steel-concrete slabs. Int J Steel Struct 15(2):415–432

Al-Deen S, Ranzi G, Vrcelj Z (2011) Full-scale long-term and ultimate experiments of simply-supported composite beams with steel deck. J Constr Steel Res 67(10):1658–1676

Bažant ZP (1972) Prediction of creep effects using age-adjusted effective modulus method. ACI Struct J 69(4):212–217

Bažant ZP, Baweja S (1995) Creep and shrinkage prediction model for analysis and design of concrete structures—model B3. Mater Struct 28:357–365, 415–430, 488–495

Bradford MA (2010) Generic modelling of composite steel-concrete slabs subjected to shrinkage, creep and thermal strains including partial interaction. Eng Struct 32(5):1459–1465

BS EN 1992-1-1 (2004). Eurocode 2: design of concrete structures—part 1-1: general rules and rules for buildings. British Standards Institution, European Committee for Standardization, Brussels

Carrier RE, Pu DC, Cady PD (1975) Moisture distribution in concrete bridge decks and pavements. Durability of Concrete, SP-47. American Concrete Institute, Farmington Hills, pp 169–192

Dilger W, Neville AM (1971) Method of creep analysis of structural members. ACI SP 27–17:349–379

Fielders Australia PL (2008) Specifying fielders. KingFlor; Composite Steel Formwork System Design Manual

Gilbert RI (1999) Deflection calculation for reinforced concrete structures—why we sometimes get it wrong. ACI Struct J 96(6):1027–1032

Gilbert RI, Ranzi G (2011) Time-dependent behavior of concrete structures. Spon Press, London **426 pp**

Gilbert RI, Bradford MA, Gholamhoseini A, Chang ZT (2012) Effects of shrinkage on the long-term stresses and deformations of composite concrete slabs. Eng Struct 40:9–19

Ranzi G, Ambrogi L, Al-Deen S, Uy B (2012) Long-term experiments of post-tensioned composite slabs. In: Proceedings of the 10th international conference on advances in steel concrete composite and hybrid structures, July 2012, Singapore

Ranzi G, Leoni G, Zandonini R (2013) State of the art on the time-dependent behaviour of composite steel-concrete structures. J Constr Steel Res 80:252–263

Standards Australia (2009) "Australian Standard for Concrete Structures", AS 3600-2009. Australia, Sydney

7

Wind interference effect on an octagonal plan shaped tall building due to square plan shaped tall buildings

Rony Kar[1] · Sujit Kumar Dalui[1]

Abstract The variation of pressure at the faces of the octagonal plan shaped tall building due to interference of three square plan shaped tall building of same height is analysed by computational fluid dynamics module, namely ANSYS CFX for 0° wind incidence angle only. All the buildings are closely spaced (distance between two buildings varies from 0.4h to 2h, where h is the height of the building). Different cases depending upon the various positions of the square plan shaped buildings are analysed and compared with the octagonal plan shaped building in isolated condition. The comparison is presented in the form of interference factors (IF) and IF contours. Abnormal pressure distribution is observed in some cases. Shielding and channelling effect on the octagonal plan shaped building due to the presence of the interfering buildings are also noted. In the interfering condition the pressure distribution at the faces of the octagonal plan shaped building is not predictable. As the distance between the principal octagonal plan shaped building and the third square plan shaped interfering building increases the behaviour of faces becomes more systematic. The coefficient of pressure (C_p) for each face of the octagonal plan shaped building in each interfering case can be easily found if we multiply the IF with the C_p in the isolated case.

Keywords Computational fluid dynamics · Interference effect · Tall building · Channelling effect · Shielding effect · Interference factor

Introduction

With the advent of latest analysis and design technology as well as high strength materials, the number of high-rise buildings is increasing. So wind engineering is getting more and more importance as the need for calculation of wind impact and interference of other structures on tall buildings arise. Wind engineering analyses the effects of wind in the natural and the built environment and studies the possible damage, inconvenience or benefits which may result from wind. In the field of structural engineering it includes strong winds, which may cause discomfort and damage, as well as extreme winds, such as in a cyclone, tornado, hurricane which may cause widespread destruction. Effect of wind on any structures is considered mainly in two directions: one is acting along the flow of wind which is called drag and the other is perpendicular to the wind flow which is called lift. Structures are subjected to aerodynamic forces which include both drag and lift. If the distance between centre of rigidity of the structure and the centre of the aerodynamic forces is large the structure is also subjected to torsional moments which may significantly affect the structural design. Interference effects due to wind are caused by the presence of adjacent structures, resulting in a change in wind loads on the principal building with respect to the isolated condition. The change mainly depends upon the shape, size and relative positions of these buildings as well as the wind incidence angle and upstream exposure. The main reasons for the lack of a comprehensive and generalized set of guidelines for wind

✉ Sujit Kumar Dalui
sujit_dalui@rediffmail.com

Rony Kar
ronykar123@gmail.com

[1] Department of Civil Engineering, Indian Institute of Engineering Science and Technology, Shibpur, Howrah, India

load modifications caused by interference effects are as follows. There are a large number of variables involved including the height and plan shape and size of buildings, their distances from one another, wind incidence angles, topographical condition and different meteorological condition. There is a general misconception that the severity of wind loads on a building is less if surrounded by other structures than the isolated condition. Many works done earlier in the field of wind engineering include wind pressure characteristics, wind flow, dynamic response, interference effect etc. for tall structures as well as short structures. Cheng et al. (2002) performed aeroelastic model tests to study the across wind response and aerodynamic damping of isolated square-shaped high-rise buildings. Kim and You (2002) investigated the tapering effect for reducing wind-induced responses of a tapered tall building, by conducting high-frequency force-balance test. Lakshmanan et al. (2002) conducted pressure measurement studies on models of three structures with different plan shapes—a circular, an octagonal and an irregular shape under simulated open terrain conditions using the boundary layer wind tunnel. Thepmongkorna et al. (2002) investigated interference effects from neighbouring buildings on wind-induced coupled translational–torsional motion of tall buildings through a series of wind tunnel aeroelastic model tests. Tang and Kwok (2004) conducted a comprehensive wind tunnel test program to investigate interference excitation mechanisms on translational and torsional responses of an identical pair of tall buildings. Xie and Gu (2004) tried to find the mean interference effects between two and among three tall buildings studied by a series of wind tunnel tests. Both the shielding and channelling effects are discussed to understand the complexity of the multiple-building effects. Lam et al. (2008) investigated interference effects on a row of square-plan tall buildings arranged in close proximity with wind tunnel experiments. Agarwal et al. (2012) conducted a comprehensive wind tunnel test programme to investigate interference effects between two tall rectangular buildings. Tanaka et al. (2012) carried out a series of wind tunnel experiments to determine aerodynamic forces and wind pressures acting on square-plan tall building models with various configurations: corner cut, setbacks, helical and so on. The number of literatures on interference effect on tall buildings other than rectangular plan shape is quite low. Further the wind loading codes [(ASCE 7–10, IS-875 (Part 3) 1987, AS/NZS: 1170.2:2002, etc.] does not provide any guidelines for incorporating interference effect in structural design, which necessitates more research on this area. The current work mainly focuses on the wind-induced interference effect on an octagonal plan shaped tall building due to the presence of three square plan shaped building of same height for 0° wind incidence angle.

Numerical analysis of a tall building by CFD

In the present study the octagonal plan shaped building in isolated as well as interference condition is analysed by the CFD package namely ANSYS CFX (version 14.5). The boundary layer wind profile is governed by the power law equation:

$$U(z) = U_\infty \left(\frac{z}{z_0}\right)^\alpha$$

A power law exponent of 0.133 is used which satisfies terrain category II as mentioned in IS 875-part III (1987).

Details of model

The buildings are modelled in 1:300 scale and the wind velocity scale is 1:5 (scaled down velocity 10 m/s). K-ε model is used for the numerical simulation. The k-ε models use the gradient diffusion hypothesis to relate the Reynolds stresses to the mean velocity gradients and the turbulent viscosity. The turbulent viscosity is modelled as the product of a turbulent velocity and turbulent length scale. k is the turbulence kinetic energy and is defined as the variance of the fluctuations in velocity. It has dimensions of $(L^2\ T^{-2})$. ε is the turbulence Eddy dissipation and has dimensions of per unit time. The continuity equation and momentum equations are:

$$\frac{\partial \rho}{\partial t} + \frac{\partial}{\partial xj}(\rho U j) = 0 \tag{1}$$

$$\frac{\partial \rho U_i}{\partial t} + \frac{\partial}{\partial x_j}(\rho U_i U_j) = -\frac{\partial p'}{\partial x_i} + \frac{\partial}{\partial x_j}\left[\mu_{\text{eff}}\left(\frac{\partial U_i}{\partial x_j} + \frac{\partial U_j}{\partial x_i}\right)\right] + S_M \tag{2}$$

where S_M is the sum of body forces, μ_{eff} is the effective viscosity accounting for turbulence, and p' is the modified pressure. ρ and U denote density and velocity respectively. The k-ε model is based on the eddy viscosity concept, so that

$$\mu_{\text{eff}} = \mu + \mu_t \tag{3}$$

μ_t is the turbulence viscosity.

$$\mu_t = C_\mu \rho \frac{k^2}{\varepsilon} \tag{4}$$

The values of k and ε come directly from the differential transport equations for the turbulence kinetic energy and turbulence dissipation rate:

$$\frac{\partial(\rho k)}{\partial t} + \frac{\partial}{\partial x_j}(\rho k U_j)$$
$$= \frac{\partial}{\partial x_j}\left[\left(\mu + \frac{\mu_t}{\sigma_k}\right)\frac{\partial k}{\partial x_j}\right] + P_k + P_{\text{b}} - \rho\varepsilon - Y_M + S_k \tag{5}$$

$$\frac{\partial(\rho\varepsilon)}{\partial t} + \frac{\partial}{\partial x_j}(\rho\varepsilon U_j) = \frac{\partial}{\partial x_j}\left[\left(\mu + \frac{\mu_t}{\sigma_\varepsilon}\right)\frac{\partial\varepsilon}{\partial x_j}\right] + \rho C_1 S_\varepsilon$$
$$- \rho C_2 \frac{\varepsilon^2}{k + \sqrt{v\varepsilon}} + C_{1\varepsilon}\frac{\varepsilon}{k}C_{3\varepsilon}P_b + S_\varepsilon$$

$$(6)$$

P_k represents the generation of turbulence kinetic energy due to the mean velocity gradients, P_b is the generation of turbulence kinetic energy due to buoyancy and Y_m represents the contribution of the fluctuating dilatation in compressible turbulence to the overall dissipation rate, C_1 and C_2 are constants. σ_k and σ_ε are the turbulent Prandtl numbers for k (turbulence kinetic energy) and ε (dissipation rate). The values considered for $C_{1\varepsilon}$, σ_k and σ_ε are 1.44, 1 and 1.2 respectively.

Domain and meshing

A domain having $5h$ upwind fetch, $15h$ downwind fetch, $5h$ top and side clearance, where h is the height of the model as shown in Fig. 1 is constructed as per recommendation of Franke et al. (2004). Such a large domain is enough for generation of vortex on the leeward side and avoids backflow of wind. Moreover no blockage correction is required. Tetrahedral elements are used for meshing the domain. The mesh near the building is smaller compared to other location so as to accurately resolve the higher gradient region of the fluid flow. The mesh inflation is provided near the boundaries to avoid any unusual flow.

The velocity of wind at inlet is 10 m/s. No slip wall is considered for building faces and free slip wall for top and side faces of the domain. The relative pressure at outlet 0 Pa. The operating pressure in the domain is considered as 1 atm, i.e. 101,325 Pa. The Reynolds number of the model varies from 3.7×10^6 to 4.0×10^6.

Validation

Before starting the numerical analysis of the octagonal plan shaped building the validity of the ANSYS CFX package is checked. For this reason a square plan shaped building (Fig. 2) of dimension 100 mm × 100 mm and height 500 mm (i.e. aspect ratio 1:5) is analysed in the aforementioned domain by K-ε model using ANSYS CFX under uniform wind flow.

Uniform wind flow of velocity 10 m/s is provided at the inlet. The domain is constructed as per recommendation of Franke et al. (2004) as mentioned before. The face average values of coefficient of pressures are determined by ANSYS CFX package and compared with wind action codes from different countries.

From Table 1 it can be seen that result found by the package is approximately same with the face average value of

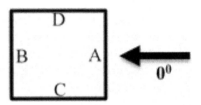

Fig. 2 Different faces of the model with direction of wind

Table 1 Comparison of face average values of coefficients of pressure

Wind loading code	Face-A	Face-B	Face-C	Face-D
By ANSYS CFX	0.83	−0.47	−0.6	−0.6
ASCE 7–10	0.8	−0.5	−0.7	−0.7
AS/NZS-1170.2(2002)	0.8	−0.5	−0.65	−0.65
IS: 875 (part 3) (1987)	0.8	−0.25	−0.8	−0.8

Fig. 1 Computational domain for isolated building model used for CFD simulation

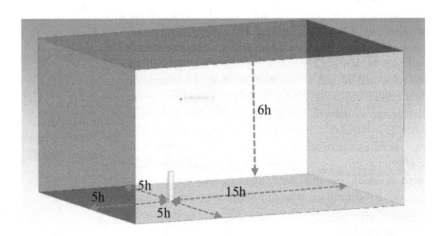

coefficient of pressure mentioned in *AS/NZS-1170.2(2002)*. While there is difference in other codes it is due to different flow conditions and methods adopted.

Parametric study

The actual height of the building is 150 m and the diameter of the circle inscribed in the plan shape is 30 m for both the principal octagonal plan shaped building and square plan shaped interfering buildings. The buildings are modelled in 1:300 scale. The scaled down height of the buildings is $h = 500$ mm and the scaled down diameter of the circle inscribed in the plan shape is 100 mm for all the buildings. The aspect ratio is 1:5 for the principal as well as interfering buildings. The numerical simulation is carried out for the octagonal plan shaped building in the presence of three square plan shaped buildings as shown in Fig. 3 for *0° wind incidence angle* only. The spacing from the principal building to upstream interfering buildings, i.e. $S_1 = 200$ mm (=$0.4h$). The spacing between the upstream interfering buildings, i.e. S_2 is varied. The spacing between

the principal and the third interfering building, i.e. S_3 is also varied.

Results and discussion

Isolated condition

In this case the octagonal plan shaped building is subjected to boundary layer wind flow at 0° wind incidence angle and analysed using k-ε turbulence model by ANSYS CFX.

Flow pattern

The flow pattern around the building is shown in Fig. 4.

The key features observed from the flow pattern are summarized below

1. As the plan shape is symmetrical the flow pattern is also symmetrical till the formation of vortices. Thus symmetrical faces will have identical or at least similar pressure distribution.

Fig. 3 Plan view of principal and interfering buildings with direction of wind flow

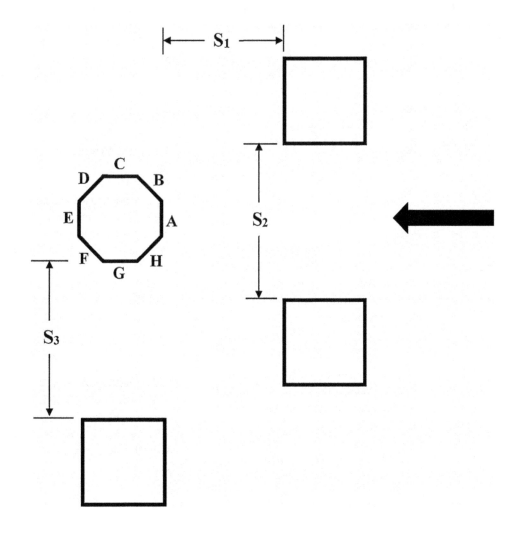

Fig. 4 Plan of flow pattern around model for 0° wind incidence angle for *k-ε* model

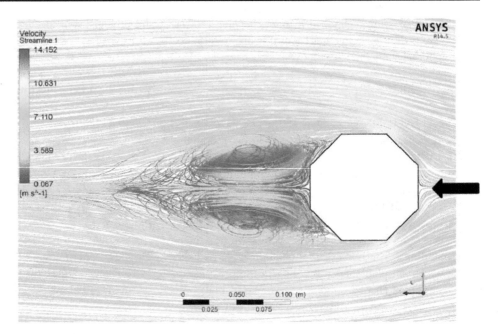

2. The wind flow separates after colliding with the windward face, i.e. face A so it will have positive pressure with slight negative values near the edges due to flow separation.
3. The inclined windward faces, i.e. faces B and H will experience slight positive value of pressure near face A junction and gradually develops negative pressure away from it.
4. The side faces C and G will have negative pressure due to side wash.
5. The back faces, i.e. faces D, E and F will experience negative pressure due to side wash and formation of vortices.

Pressure variation

The model has a symmetrical plan shape and flow pattern is also symmetrical; so only five faces are sufficient for understanding the behaviour of the model under wind action. Pressure contours of the faces A, B, C, D and E are shown in Fig. 5.

The features of the pressure contours are as follows:

1. Face A experiences mainly positive pressure except near the top edge. Pressure distribution is parabolic in nature due to boundary layer flow and symmetrical about vertical centreline.
2. Face B and H have slightly positive pressure near the junction of face A and negative elsewhere.

3. Faces C and G have throughout negative pressure with more negative value towards the leeward side.
4. Faces D and F have lower negative value at bottom and higher value towards top.
5. Face E has a semi-circular zone of lower negative value and an elliptical zone at the middle.

Face average value of pressure coefficient

Interfering condition

In this case also the buildings are subjected to a boundary layer wind flow at the wind incidence angle 0° only. The plan view of the principal octagonal as well as the interfering buildings are shown in Fig. 3. The distance between principal to front interfering buildings $S_1 = 200$ mm. The distance between principal and the third interfering building S_3 varies as 200 mm (=0.4h), 300 mm (=0.6h) and 500 mm (=h). The distance between front interfering buildings varies as $S_2 = 200$ mm (=0.4h), 500 mm (=h), 1000 mm (=2h).

Flow pattern

The wind flow pattern for different interference conditions will be different. Wind flow pattern will change as distances between principal to interfering or between interfering buildings change. In Fig. 6 the flow pattern for a typical case where $S_1 = 0.4h$, $S_2 = 0.4h$ and $S_3 = 0.4h$ is depicted.

Fig. 5 Pressure contour on different faces of Octagonal plan shaped building for 0° wind angle by k-ε method

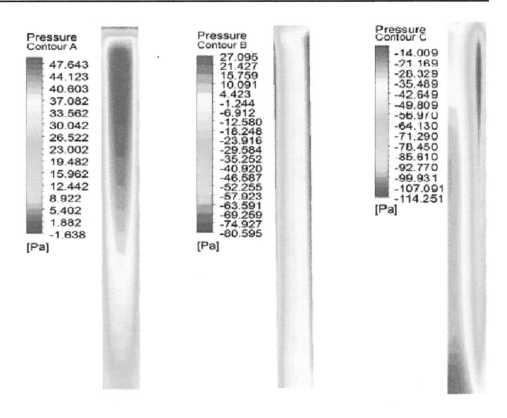

Pressure
Contour A

47.643
44.123
40.603
37.082
33.562
30.042
26.522
23.002
19.482
15.962
12.442
8.922
5.402
1.882
-1.638
[Pa]

Pressure
Contour B

27.095
21.427
15.759
10.091
4.423
-1.244
-6.912
-12.580
-18.248
-23.916
-29.584
-35.252
-40.920
-46.587
-52.255
-57.923
-63.591
-69.259
-74.927
-80.595
[Pa]

Pressure
Contour C

-14.009
-21.169
-28.329
-35.489
-42.649
-49.809
-56.970
-64.130
-71.290
-78.450
-85.610
-92.770
-99.931
-107.091
-114.251
[Pa]

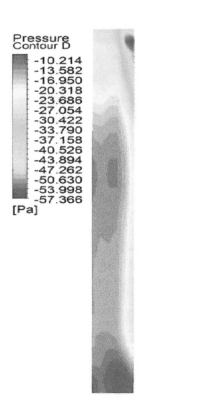

Pressure
Contour D

-10.214
-13.582
-16.950
-20.318
-23.686
-27.054
-30.422
-33.790
-37.158
-40.526
-43.894
-47.262
-50.630
-53.998
-57.366
[Pa]

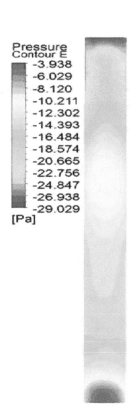

Pressure
Contour E

-3.938
-6.029
-8.120
-10.211
-12.302
-14.393
-16.484
-18.574
-20.665
-22.756
-24.847
-26.938
-29.029
[Pa]

Fig. 6 Plan of flow pattern around building setup for $S_1 = 0.4h$, $S_3 = 0.4h$

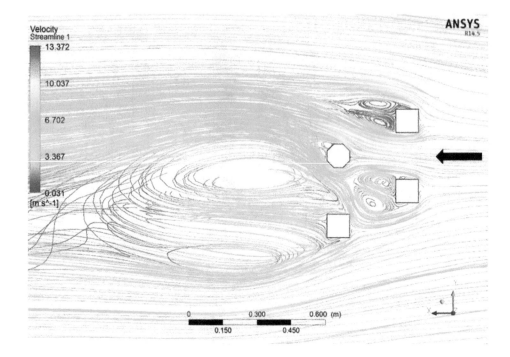

The features observed are as follows:

1. Unsymmetrical vortices are formed in the leeward side of the principal building due to interference effects of the other interfering buildings.
2. Pressure in the windward face is almost same. Maybe due to small distance between principal and front buildings is very so less that channelling effect of the front interfering buildings does not affect the principal building.
3. Faces B, C, D, E and F have higher magnitude of pressure due to both channelling effect and flow separation.
4. Surprisingly faces F, G and H have lower pressure probably due to Shielding effect of the third interfering building.

Interference factor contour

Interference factor for selected points is given by,

$$\text{IF}_p = \frac{\text{Pressure for selected point in interfering condition}}{\text{Pressure for selected point in isolated condition}}$$

(7)

Interference factor (IF) thus found can be used to plot interference factor contour for each face. Interference factor for above mentioned case for faces A, C, E, F and H are shown in Fig. 7.

The key features observed from the IF contours are as follows:

1. There is no interference in face A except for the top portion of the face.
2. Faces C and F has interference throughout the surfaces.
3. Face E seems to have negligible interference slightly below the horizontal centreline of the face.
4. Face H also has negligible interference except for the region slightly below the top portion.

Interference factor contour can be a pretty useful tool for depicting the local interference of a building face.

Case I: $S_1 = 0.4h$, $S_3 = 0.4h$

In this case S_1 and S_3 remains constant and S_2 varies as $0.4h$, h and $2h$. The comparison of pressure coefficient (C_p) on all faces, i.e. A, B, C, D, E, F, G and H along the vertical centreline depicting the variation as the distance S_2 varies is shown in Fig. 8.

The inferences drawn from the plot of pressure coefficient along vertical centreline are as follows:

1. Face A does not experience much variation from each other or the isolated case. The front interfering buildings are too close to the principal building so the channelling effect does not affect the front face.
2. In case of face B the magnitude of C_p is higher when $S_2 = 0.4h$ than the other cases. The reason is the combined effect of channelling effect by the front interfering buildings and shielding effect of the third interfering building present at the other side which can

Fig. 7 Contours of interference factors for; **a** face A, **b** face C, **c** face E, **d** face F, **e** face H

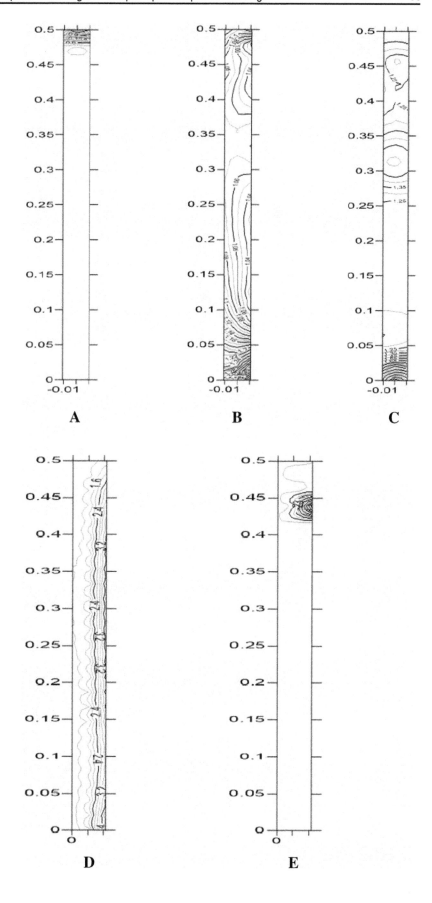

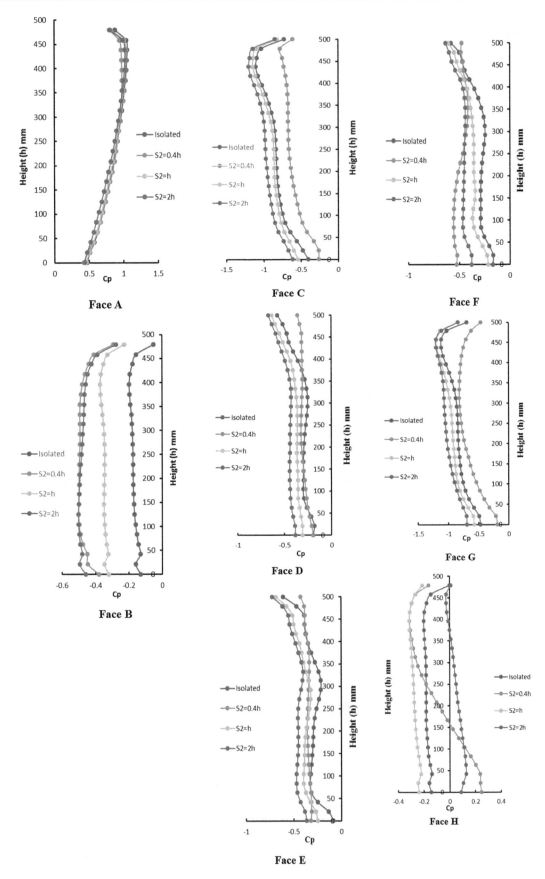

Fig. 8 Comparison of variation of pressure coefficient along vertical centreline if $S_3 = 0.4h$ for different faces

be proved by the higher velocity of wind in the zone near the faces B, C and D.

3. For the faces C, D, E and G the general trend is as S_2 increases the magnitude of C_p also increases.
4. The face F also behaves like the face B, i.e. magnitude of C_p is much higher at $S_2 = 0.4H$ than the other cases.
5. The behaviour of face H is completely arbitrary except for the case $S_2 = 2H$ where the C_p have positive values deviating from the usual behaviour. The reason maybe that the channelling effect of the upstream interfering buildings becomes negligible and the shielding effect of the third interfering building becomes dominant.

It can be observed that the behaviour is pretty haphazard except for some cases as described previously.

Case II: $S_1 = 0.4h$, $S_3 = 0.6h$

In this case S_1 and S_3 remains constant and S_2 varies as 0.4H, H and 2H. The comparison of pressure coefficient (C_p) on all faces, i.e. A, B, C, D, E, F, G and H along the vertical centreline depicting the variation as the distance S_2 varies is shown in Fig. 9.

The key features observed from the plot of pressure coefficient along vertical centreline are as follows:

1. Like the previous case in this case also the variation in the plot of C_p along vertical centreline for different cases is not so prominent due to the similar reason.
2. In this case face B experiences lesser pressure as S_2 increases. The cause is the decrease in velocity when compared to the previous case due to the increase in the distance S_3.
3. For the faces C, D, E, F and G the magnitude of pressure coefficient increases when the distance S_2 increases from 0.4h to 0.6h which is due to the decrease of channelling effect caused by the upstream interfering buildings. The pressure coefficient again decreases when $S_2 = h$ probably due to decrease in interference by the upstream buildings on the principal building.
4. In this case face H exhibits steady decrease in pressure coefficient due to the shielding effect of the third interfering building at the side. The magnitude of pressure coefficient is even lower than the value at isolated condition.

In this case it can be seen that the variation of pressure coefficient is pretty systematic when compared to the previous case.

Case III: $S_1 = 0.4h$, $S_3 = h$

In this case S_1 and S_3 remains constant and S_2 varies as 0.4h, h and 2h. The comparison of pressure coefficient (C_p)

on all faces, i.e. A, B, C, D, E, F, G and H along the vertical centreline depicting the variation as the distance S_2 varies is shown in Fig. 10.

The key features observed from the plot of pressure coefficient along vertical centreline are as follows:

1. Similar to the previous two cases the plot of pressure coefficient along height does not vary much when S_2 increases.
2. Face B experiences less pressure coefficient as the distance S_2 increases. The magnitude of pressure coefficient is lesser than that in the previous case. The cause is the velocity of wind in that region decreases as compared to the previous case due increase in the distance S_3, i.e. distance between principal building and the third interfering building.
3. For the faces C, D, E, F and G the magnitude of pressure coefficient increases as the distance S_2 increases. In this case the magnitudes of pressure coefficients at $S_2 = 0.6h$ and $S_2 = h$ are almost same for these faces.
4. For face H also similar trend is observed, i.e. magnitude of pressure coefficient decreases with the increase of S_2. But here the magnitudes of pressure coefficients are greater than that of the isolated case. The behaviour deviates from the previous case from which it can be concluded that when the distance from the third building from the principal building, i.e. $S_3 = h$, the shielding effect of the third building reduces substantially.

This case is pretty different from the previous two cases as the effect of third building on the principal building is reduced.

Interference factor

Interference effects are presented in the form of non-dimensional interference factors (IF) that represent the aerodynamic forces on an octagonal plan shaped principal building with interference from adjacent three square plan shaped buildings. IF is given by the following formula

$$\text{I.F.} = \frac{\text{Mean pressure for a face in interfering condition}}{\text{Mean pressure for a face in isolated condition}}$$
(8)

Here is a guideline for wind load modifications in planning and designing an octagonal plan shaped building surrounded by some square plan shaped buildings. If C_p be the face average value of pressure coefficient for a particular face in isolated condition then the same for any particular interfering condition is given by

$$C_{p,\text{interfering}} = \text{I.F.} \times C_{p,\text{isolated}}$$
(9)

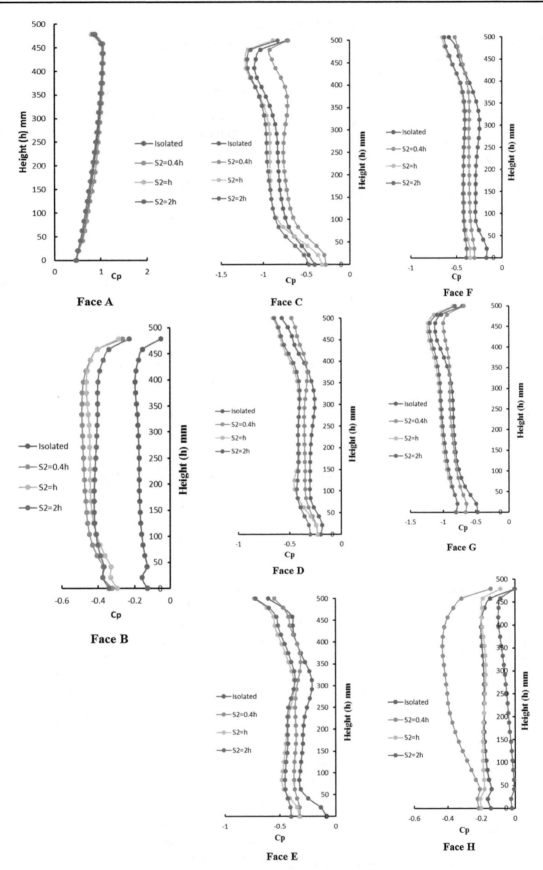

Fig. 9 Comparison of variation of pressure coefficient along vertical centreline if $S_3 = 0.6h$ for different faces

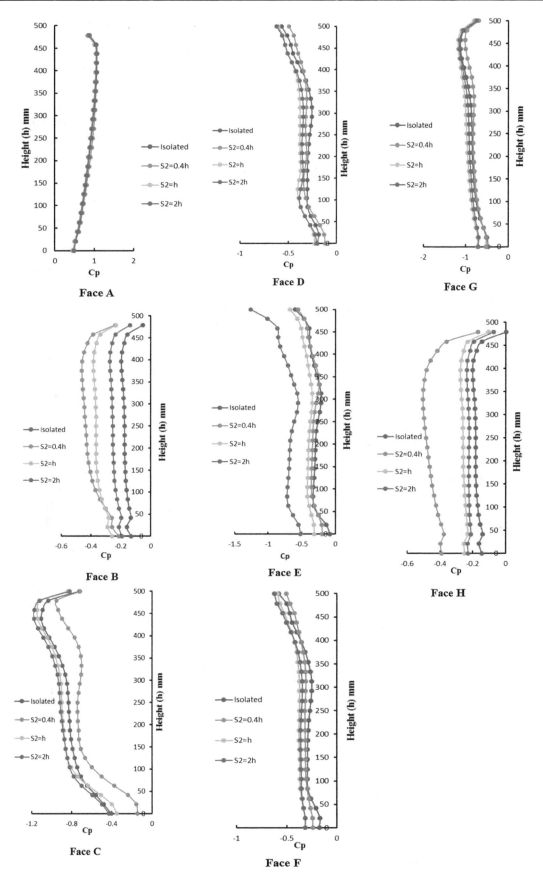

Fig. 10 Comparison of variation of pressure coefficient along vertical centreline if $S_3 = h$ for different faces

Table 2 Face average values of pressure coefficients for 0° wind incidence angle

Location	Face average values of pressure coefficient
Face A	0.78
Face B	−0.29
Face C	−0.94
Face D	−0.37
Face E	−0.58
Face F	−0.36
Face G	−0.92
Face H	−0.29

where IF is the interference factor of that face in that particular condition. $C_{p,\text{isolated}}$ can be found from Table 2.

The interference factor for the previously mentioned cases are tabulated in Table 3. The interference factor for the intermediate cases can be calculated by linear interpolation.

Conclusion

The study carried out till now has shown regular plan shape buildings experiences symmetrical pressure distribution for 0° wind incidence angle in isolated condition. The interference effect on different faces should also be kept in mind while calculating the wind load on buildings or any other important structures. The significant outcomes of this present study on the "octagonal" plan shaped tall building can be summarized as follows:

1. The octagonal plan shaped building experiences symmetrical pressure distribution in isolated condition.
2. In the interfering condition the pressure distribution cannot be predicted accurately. It can be seen from

Table 3 that no particular pattern in IF can be seen with the change in distances between the buildings.
3. When the distance between the principal building and the third interfering building is 200 mm (i.e. $S_3 = 0.4h$) the behaviour cannot be predicted with the change in S_2 with some exception. But the channelling effect of the two upstream buildings and the shielding effect of the third interfering building is evident.
4. In the case II where the distance between the principal building and the third interfering building is 300 mm (i.e. $S_3 = 0.6h$) the behaviour of the faces due to interference becomes slightly more systematic than the previous case. As the distance between two upstream buildings increase the shielding effect of the third interfering building becomes more prominent.
5. In the case III where the distance between the principal building and the third interfering building is 500 mm (i.e. $S_3 = h$) the behaviour of the faces due to interference remains similar to the previous case. The difference is that S_3 is increased so the channelling effect of the upstream interfering buildings become prevalent and the shielding effect due to the third interfering building decreases.
6. From Table 3, we can see that in many occasions the values of the interference factor is greater than unity, i.e. the coefficient of pressure in that particular interfering case is greater than that in isolated case. This proves that the presence of the interfering buildings does not always contribute to the decrease of wind load on the principal building.
7. The pressure coefficients for the interfering cases can be easily found out from expression (9) if the same for the isolated case and the corresponding IF for the interfering case is known for the octagonal plan shaped building of the similar aspect ratio.

Table 3 Mean interference factor for each faces of principal octagonal building for different cases

S_1 (mm)	S_3 (mm)	S_2 (mm)	Interference factor for faces							
			A	B	C	D	E	F	G	H
0.4h	0.4h	0.4h	0.95	1.86	1.08	1.10	1.09	1.64	0.84	0.54
		h	0.95	1.59	1.08	1.14	1.28	1.17	1.11	0.42
		2h	0.99	2.06	1.17	1.32	1.50	1.59	1.15	0.38
	0.6h	0.4h	0.96	1.80	0.92	0.95	1.18	1.16	1.03	1.46
		h	0.95	1.85	1.11	1.23	1.46	1.38	1.19	1.12
		2h	0.99	1.03	1.22	1.30	1.24	1.20	1.15	0.77
	h	0.4h	0.95	1.61	0.88	0.87	1.05	0.98	0.99	1.77
		H	0.96	1.61	1.07	1.13	1.31	1.23	1.13	1.30
		2h	0.99	1.25	1.07	1.15	1.27	1.20	1.10	1.22

References

Agarwal N, Mittal AK, Gupta VK (2012) Along-wind interference effects on tall buildings. In: National Conference on Wind Engineering, pp 193–204

Cheng CM, Lu PC, Tsai MS (2002) Acrosswind aerodynamic damping of isolated square shaped buildings. J Wind Eng Ind Aerodyn 90:1743–1756

Franke J, Hirsch C, Jensen A, Krüs H, Schatzmann M, Westbury P, Miles S, Wisse J, Wright NG (2004) Recommendations on the use of CFD in Wind Engineering. In: COST Action C14: Impact of Wind and Storm on City Life and Built Environment, von Karman Institute for Fluid Dynamics

IS 875 (Part 3)-1987 Indian Standard code of practice for design loads (other than earthquake) for building and structures, bureau of indian standard, New Delhi

Kim YM, You KP (2002) Dynamic responses of a tapered tall building to wind loads. J Wind Eng Ind Aerodyn 90:1771–1782

Lakshmanan N, Arunachalam S, Rajan SS, Babu GR, Shanmugasundaram J (2002) Correlations of aerodynamic pressures for prediction of acrosswind response of structures. J Wind Eng Ind Aerodyn 90:941–960

Lam KM, Leung MYH, Zhao JG (2008) Interference effects on wind loading of a row of closely spaced tall buildings. J Wind Eng Ind Aerodyn 96:562–583

Tanaka H, Tamura Y, Ohtake K, Nakai M, Kim YC (2012) Experimental investigation of aerodynamic forces and wind pressures acting on tall buildings with various unconventional configurations. J Wind Eng Ind Aerodyn 107–108:179–191

Tang UF, Kwok KCS (2004) Interference excitation mechanisms on a 3DOF aeroelastic CAARC building model. J Wind Eng Ind Aerodyn 92:1299–1314

Thepmongkorna S, Wood GS, Kwok KCS (2002) Interference effects on wind-induced coupled motion of a tall building. J Wind Eng Ind Aerodyn 90:1807–1815

Xie ZN, Gu M (2004) Mean interference effects among tall buildings. Int J Eng Struct 26:1173–1183

8

Seismic response of an elevated aqueduct considering hydrodynamic and soil-structure interactions

Bhavana Valeti[1] · Samit Ray-Chaudhuri[1] · Prishati Raychowdhury[1]

Abstract In conventional design of an elevated aqueduct, apart from considering the weight of water inside the channels, hydrodynamic forces are generally neglected. In a few special cases involving high seismic zones, hydrodynamic forces have been modeled considering equivalent lumped-mass type idealization or other models. For support conditions, either the base is considered as fixed or in a few cases, equivalent spring-dashpot system is considered. However, during an intense seismic event, nonlinear soil-structure interactions (SSI) may alter the response of the aqueduct significantly. This paper investigates the effect of hydrodynamic forces and SSI on seismic response of a representative elevated aqueduct model. Different modeling concepts of SSI has been adopted and the responses are compared. Frequency domain stochastic response analysis as well as time-history analysis with a series of ground motions of varying hazard levels have been performed. Demand parameters such as base shear and drift ratio are studied for varying heights of water in channels and different site conditions. From the frequency domain analysis, the effect of convective masses is found to be significant. From the time history analysis, the overall effect of increase in height of water is found to be negligible for nonlinear base case unlike the fixed and elastic base cases. For the nonlinear base condition, the base shear demand is found to decrease and the drift ratio is found to increase when compared to the results of linear base condition. The results of this study provide a better understanding of seismic behavior of an elevated aqueduct under various modeling assumptions and input excitations.

Keywords Elevated aqueduct · Seismic response · Hydrodynamic effects · Soil-structure interaction · Stochastic response · Nonlinear time-history analysis

Introduction

Aqueducts are man made structures that have been crucial part of every civilization in the distribution of the essential but not so ubiquitous element of nature, water. Aqueducts transport water across topographical barriers to their destination, taking various forms and traveling at different levels with respect to the ground such as pipelines and canals. An elevated aqueduct is a bridge that crosses barriers such as a valley or a river, often accommodating road and water transport systems.

The massive load of water in an elevated aqueduct shifts its center of mass further above the ground compared to highway/railway bridges. As a result, the structure becomes more vulnerable to dynamic lateral forces, especially, those due to hydrodynamic effects. In addition, for estimation of seismic response, a fixed base assumption for stiff structures such as an aqueduct, may often lead to an inappropriate design if the soil underneath is not so stiff. Consideration of nonlinear SSI in such a situation can yield a more realistic appraisal of the behavior of soil-foundation interface during strong earthquakes. This is because the nonlinear SSI can take into account the energy dissipation behavior and nonlinear variation of stiffness along the soil-foundation interface.

(a) *Hydrodynamic forces* A water bearing structure experiences hydrodynamic forces when subjected to inertial forces. Such dynamic forces were first taken into account by Westergaard (1933) for determining the dynamic pressures due to water on a rectangular

✉ Prishati Raychowdhury
prishati@iitk.ac.in

[1] Indian Institute of Technology Kanpur, Kanpur, India

dam with a vertical face. Perhaps Housner (1954) was the first to propose a method to model dynamic effects of fluid in accelerated containers in terms of convective and impulsive pressures. Liu (1981) applied a Lagrangian-Eulerian method for the kinematical description of fluid-structure interaction. Ramaswamy and Kawahara (1987) analyzed large free surface motions in the fluid domain including sloshing using an arbitrary Lagrangian-Eulerian kinematical description. A particle finite element method was used by Idelsohn et al. (2006) to discretize the fluid continuum into particles. A number of studies were carried out on the hydrodynamic effects of ground supported water tanks. Details of these studies can be found in Valeti (2013).

(b) *SSI effects* In current practice, although nonlinear design of superstructure is accepted to meet its ductility demand, the foundations are generally designed to remain in linear zone. This is because of the difficulty in inspection and repair of highly deformed foundations along with the concern in reliable estimation of nonlinear soil-foundation responses. Gutierrez and Chopra (1978) evaluated the methods for seismic analysis of SSI, namely, simple general substructure methods and direct finite element method. Impedance functions for horizontal and coupled degrees of freedom were developed for soil modeled as both uniform viscous medium and layered soil over uniform viscous medium (Wong and Luco 1985; Gazetas 1991b). Harden et al. (2005) calibrated the model parameters for the Beam-on-nonlinear-winkler-foundation (BNWF) model. This model was subsequently updated by Raychowdhury and Hutchinson (2009) and a command *ShallowFoundationGen* was introduced in the OpenSees (2012) platform. Raychowdhury (2011) and Raychowdhury and Singh (2012) studied the effects of soil compliance and nonlinearity in low-rise steel moment-resisting frames subjected to earthquake ground motions of varying hazard levels. It is clear from the aforementioned discussions that SSI and hydrodynamic forces may be crucial for performance assessment of elevated aqueducts under significant earthquake loading. However, as per authors' knowledge, not much work has been done so far considering these aspects. The present study addresses these two important issues to develop a better understanding of seismic behavior of an elevated aqueduct under various modeling assumptions and input excitations. For this purpose, a simplified model of an elevated aqueduct is considered. Different base fixity conditions such as fixed-base, elastic-base and nonlinear base conditions are

considered. Housner's model is used to represent water in aqueduct channels for seismic analysis. For nonlinear SSI model, the 'Beam-on-Nonlinear-Winkler-Foundation (BNWF)' concept is used. Time history analyses are performed with a series of ground motions of varying hazard levels to study the effect of nonlinear SSI. Demand parameters such as base shear and drift ratio are studied for varying heights of water in channels and different soil conditions at the site.

Numerical modeling

Seismic analysis of an elevated aqueduct involves developing appropriate models to represent: (1) structural model of the aqueduct, (2) hydrodynamic effects between water and aqueduct, and (3) SSI effects between soil and foundation. The following subsections provide a detailed discussion of the numerical modeling approach for the aforementioned components:

The superstructure of an elevated aqueduct comprises the deck slab and walls, which in turn rest on substructure, i.e., on piers and abutments. Bearings are used between the deck and substructure to restrict any vertical or transverse relative movement of the deck. Only limited movement is allowed in the longitudinal direction to accommodate thermal expansions. The deck conveys water through a single or multiple channels separated by walls. The foundations underneath the pier walls can either be deep or shallow depending soil conditions. In addition to the loads acting on a typical (railway/highway) bridge, forces due to sloshing and impact of water inside the channels act laterally. This makes an aqueduct more vulnerable to seismic forces compared to (railway/highway) bridges.

In this study, Arjun Feeder Canal Aqueduct proposed to be constructed across Birma river located in the state of Madhya Pradesh (India) is representatively modeled. The aqueduct considered is 188.4 m long from abutment to abutment with 13 spans each 14.5 m long. The deck is 31.5 m wide with a slab of thickness 0.65 m and is equally divided into four channels, each of width (L) 6.938 m. The channels are separated by 3.11 m high walls of 0.75 m thickness each. The deck is at an average height of (H_a) 15 m. The pier walls are 34.5 m long (L_p) and 1.732 m thick (B_p). The foundation is a raft connecting all the pier walls between the abutments (see Fig. 1).

For simplicity of modeling, only a representative portion of the aqueduct is modeled as a single-degree of freedom (SDOF) system. For this purpose, a tributary span of the deck on a single pier is idealized as a classically damped, linear, SDOF system with mass m_p concentrated at an effective height H_p as shown in Fig. 1. The mass m_p comprises of (1) mass of deck, (2) impulsive mass of water and (3) mass from top half portion of the pier wall. The stiffness of the superstructure (K_p) is calculated from the stiffness of the pier. The pier wall is numerically modeled using *ElasticBeamColumnElement* in OpenSees (2012). A damping ratio of 5 % is used for modeling of the aqueduct structure.

In Fig. 1, also shown are the convective water masses connected with representative springs. In this study, Housner's model (Housner 1954) (see Fig. 2) is employed in the calculation and modeling of the dynamic pressure distribution in terms of convective and impulsive masses acting on the walls of a channel of the aqueduct. For unit length of the aqueduct, the equivalent impulsive mass M_o due to static pressure distribution in a channel is represented as mass rigidly attached to the walls at a height H_o. Housner (1954) defined impulsive mass according to Eqs. (1) and (2). Here, H is the height of water in the aqueduct channel, L is half of the width of the channel and M is the mass of water per unit length of the channel.

$$H_o = \frac{3}{8}H \tag{1}$$

$$M_o = M \frac{\tanh(\sqrt{3}L/H)}{\sqrt{3}L/H} \tag{2}$$

For unit length of the aqueduct, the sloshing effect of water in channel is represented by an equivalent convective water mass M_n (Housner 1954). This mass is assumed to be attached to the walls at a height H_n with stiffness K_n as given in Eqs. (3)–(5).

$$M_n = M \frac{1}{3} \sqrt{\frac{5}{2}\frac{L}{H}} \tanh\left(\sqrt{\frac{5}{2}\frac{H}{L}}\right) \tag{3}$$

$$H_n = H \left[1 - \frac{1}{\sqrt{\frac{5}{2}\frac{H}{L}} \tanh\sqrt{\frac{5}{2}\frac{H}{L}}} + \frac{1}{\sqrt{\frac{5}{2}\frac{H}{L}} \sinh\sqrt{\frac{5}{2}\frac{H}{L}}} \right] \tag{4}$$

$$K_n = (\omega_n^2)M_n = \frac{3gHM_n^2}{L^2 M} \tag{5}$$

For modeling of the hydrodynamic effects, separate impulsive and convective masses are computed for each channel of the aqueduct for unit length. These masses are then computed for the tributary length of the aqueduct channels. The walls of the aqueduct are assumed to be rigid and the static pressure acting on the bottom of the aqueduct channels are neglected. Figure 3 shows modeling of impulsive and sloshing masses for the aqueduct channels. Here, in each channel, m_o represents the impulsive mass, m_{si} is the convective mass and k_{si} is the total stiffness of springs connecting the convective mass with the channel structure. These impulsive masses are then added to the mass of the aqueduct while equivalent convective masses are fixed on top of the structure with a uniaxial springs in the transverse direction of the aqueduct (see Fig. 1). Housner's parameters for different heights of water in each channel of the aqueduct for a unit length are as shown in Table 1 for increasing heights of water up to 3 m. In addition to these, a damper with damping ratio of 0.1 % is

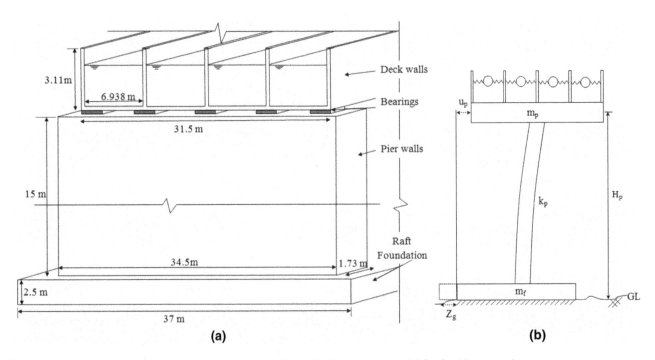

Fig. 1 **a** Schematic diagram of aqueduct in transverse direction and **b** lumped mass model for fixed-base aqueduct

Fig. 2 **a** Sloshing and impulsive masses of water and **b** Housner's model representation

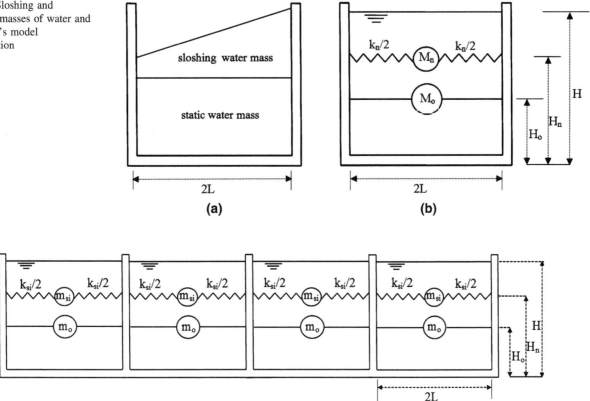

Fig. 3 Housner's representation of hydrodynamic effects in aqueduct channels

Table 1 Parameters for Housner's model

H (m)	M (kg)	M_o (kg)	m_{si} (kg)	k_{si} (kN/m)	H_o (m)	H_1 (m)	ω_{si} (rad/s)
0.5	25,150.25	20,603.09	20,638.30	2092.89	0.19	0.25	1.00
1.0	50,300.5	39,236.89	74,851.16	8371.48	0.38	0.51	1.38
1.5	75,450.75	54,619.35	145,044.89	18,823.56	0.56	0.78	1.63
2.0	100,601	66,389.10	214,290.49	33,322.11	0.75	1.06	1.80
2.5	125,751.3	74,881.88	272,623.22	51,473.93	0.94	1.37	1.90
3.0	150,901.5	80,754.25	317,059.08	72,648.91	1.13	1.70	1.98

added with each sloshing mass to provide additional energy dissipating mechanism during sloshing.

Modeling of SSI

The foundation for the representative portion of the structure is assumed as a shallow foundation with 14.5 m along the longitudinal direction of aqueduct (i.e, along the deck) and 34.5 m in the transverse direction (across the deck) and 1.5 m deep. The foundation is assumed to be a mat foundation resting on the ground. For the shake comparison of responses, the soil-foundation interface is modeled as fixed, elastic and nonlinear base cases.

Modeling of linear SSI for frequency domain analysis Gazetas (1991a) impedance functions are employed to model SSI in frequency domain analysis. The foundation is

assumed to be rigid, massless and placed on a homogeneous elastic half-space, which replicates a reasonably deep, uniform soil deposit (Gazetas 1991a). Since, impedance functions are dependent on soil properties, geometry of foundation and excitation frequency, they are suitable in the stochastic response analysis considering SSI in frequency domain. The impedance function (S) for a given degree of freedom can be written as:

$$S = \tilde{K} + i\omega C \tag{6}$$

where ω is excitation frequency

$$\tilde{K} = Kk(\omega) \tag{7}$$

$$i\omega C = \left(C_{rad} + \frac{2\tilde{K}}{\omega} \beta \right) \tag{8}$$

Here, K is the static stiffness for a given footing and soil parameters; $k(\omega)$ is the dynamic stiffness coefficient dependent on excitation frequency and acquired from experimental results for different foundation geometries and soil properties (Gazetas 1991b). The complex damping component $i\omega C$ is a combination of radiation damping C_{rad} and inherent material damping $\frac{2\tilde{K}}{\omega}\beta$, which are dependent on excitation frequency, and β is the material damping constant of the soil. The soil parameters that have been used for different soil conditions for the calculation of impedance functions are as given in Table 2.

Modeling of SSI for time-history analysis Modeling nonlinear SSI for shallow foundations requires a model to capture the nonlinear soil-foundation behavior such as, temporary gap formation, foundation settlement, sliding and hysteretic energy dissipation. BNWF (Harden et al 2005) is one such model implemented in OpenSees (2012) with the aforementioned attributes. In BNWF model elastic foundation is modeled with discrete finite elements defined using ElasticBeamColumnElements (OpenSees 2012). Compound nonlinear independent zero-length winkler springs are attached to these foundation elements. These are defined by nonlinear hysteretic materials (QzSimple, PySimple, TzSimple materials Boulanger et al. 1999). These nonlinear springs are a

combination of dashpots, and drag and gap elements that define the soil-foundation interaction by capturing horizontal $(p - x)$, vertical $(p - z)$, shear-sliding $(t - x)$ and moment rotation behaviors at the base of the footing. The *ShallowFoundationGen* (Raychowdhury and Hutchinson 2008) command is used to model the soil-foundation interface with lesser inputs from the user reducing the manual task of defining each element of the soil-foundation system. It uses the concept of BNWF. The command also provides options for the degree of flexibility at the soil-foundation interface as shown in Fig. 4, where k_{in} represent the initial stiffness of of foundation springs. Input parameters for foundation modeling in OpenSees using *ShallowFoundationGen* are as discussed here. The soil properties of the clayey soils considered are taken as in Table 2. The dimensions of foundation are 37 m long, 6 m wide, and 2.5 m deep. A nominal embedment depth of 0.01 m is considered for the convenience of analysis. For mesh generation, *ShallowFoundationGen* requires the input for the following factors: (a) stiffness intensity ratio (R_k) for vertical springs at the end of the footing to those at the middle, which is taken as 2, (b) end length ratio (R_e) between length of the stiffened ends to the length of the foundation, which is taken as 0.2, and (c) vertical spacing (S_e) between the springs as fraction of total footing length, which is taken as 0.2.

Table 2 Soil parameters (Bowles 1988)

Soil type	Poison'sratio (v)	Shear wave velocity $(V_s$, m/s)	Density $(\rho$, kg/m$^3)$	Elastic modulus $(E_s$, 10^6 N/m$^2)$	Cohesion $(c$, N/m$^2)$
Soft	0.2	100	1750	25	25,000
Medium	0.325	200	1950	50	75,000
Firm	0.4	300	2250	100	150,000

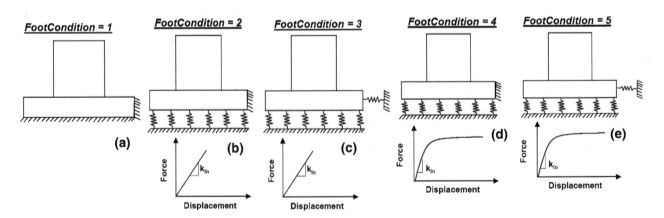

Fig. 4 Different footing conditions: **a** fixed base, **b** elastic base with no sliding allowed, **c** elastic base with sliding allowed, **d** nonlinear base with no sliding and **e** nonlinear base with sliding unrestricted (Raychowdhury and Hutchinson 2008)

Eigenvalue analysis

Eigenvalue analyses are performed for the model aqueduct for different heights of water including only the impulsive water masses. The foundation is modeled using *ShallowFoundationGen* command (OpenSees 2012) with different base fixity conditions. The natural periods for the fixed-base aqueduct structure (i.e., along the transverse direction of the aqueduct channels) are found to be very low. The increase in the periods is observed with the augmentation of flexibility due to soil at the base. For zero water height the period of the structure increased from 0.025 s for the fixed-base case to 0.52, 0.59, and 0.743 s for firm, medium and soft soil conditions, respectively. The periods have increased more than ten times with the flexibility at the base. This drastic increase in period from the fixed to flexible base case can be explained by considering the relative stiffness of the structure in comparison to the foundation springs. The stiffness of the representative model in this direction is very high due to a very long shear wall (34.5 m long and 1.732 m thick) and hence, the fixed base period is found to be very low. However, when the flexible base condition is considered, the rocking mode (due to foundation springs) influences the fundamental period significantly, even for the firm base condition. And hence, the period increases significantly even for the firm base condition. This is because, the stiffness of the foundation due to springs (even in case of firm condition) is much lower than that of the structure. To study the effect of varying water level in channels, the normalized fundamental periods of the structure for different height of water in the channels (and for the fixed and different elastic base conditions) are shown in Fig. 5. The normalization has

been done with respect to the period when the aqueduct is empty. Hence, all the curves for different base conditions start from unity. The variation of impulsive masses with height is not significant enough to notably change the period of the highly stiff aqueduct structure. A shoot up of periods of structure is not linear for any soil condition, as the increase in the impulsive mass is nonlinear and also the masses added are so small compared to the structural mass that this addition cannot lead to a linear increase in the periods.

Seismic analysis of aqueduct

This section describes the analysis procedures for the frequency domain and time-history analysis of the elevated aqueduct model considered in this study.

Frequency domain analysis

In frequency domain analysis, fixed-base and elastic base conditions for the foundation model are considered. The details of these analyses are provided as follows:

Fixed-base case Consider a linear, classically damped fixed-base SDOF structure with mass m_p, stiffness k_p and damping c_p. Let u_p denote the displacement at the top of the fixed-base structure relative to the base. Let m_{si} be the ith convective water mass attached to the aqueduct structure through springs of total stiffness k_{si} and damping coefficient c_{si}, where $i = 1, 2, \ldots, n$. Let u_{si} be the displacement of ith convective mass with respect to its base (see Fig. 6). The equations of motion for the structure and the convective water masses subjected to free-field horizontal ground acceleration, $\ddot{Z}_g(t)$ (in the transverse direction of aqueduct) can be written as follows:

For aqueduct structure,

$$m_p \ddot{u}_p(t) + c_p \dot{u}_p(t) + k_p u_p(t)$$
$$- \sum_{i=1}^{n} k_{si} u_{si}(t) - \sum_{i=1}^{n} c_{si} \dot{u}_{si}(t) = -m_p \ddot{Z}_g(t) \quad (9)$$

For water masses,

$$m_{si} \ddot{u}_{si}(t) + c_{si} \dot{u}_{si}(t) + k_{si} u_{si}(t)$$
$$= -m_{si} \ddot{Z}_g(t) - m_{si} \ddot{u}_p(t), \quad i = 1, 2, \ldots, n \quad (10)$$

Taking Fourier transform and on further simplifications, the displacement response of the aqueduct structure in frequency domain can be written as follows:

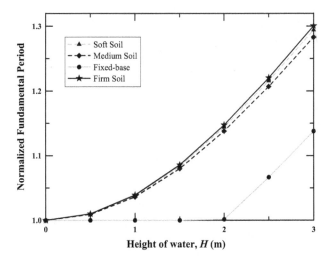

Fig. 5 Comparison of normalized natural periods of structure for different base conditions and for varying height of water

$$u_p(\omega) = \frac{H_f(\omega)\left(-1 + \frac{1}{m_p} \sum_{i=1}^{n} v_i(\omega) h_{si}(\omega)\right)}{1 + \frac{H_f(\omega)}{m_p} \omega^2 \sum_{i=1}^{n} v_i(\omega) h_{si}(\omega)} \ddot{Z}_g(\omega) \quad (11)$$

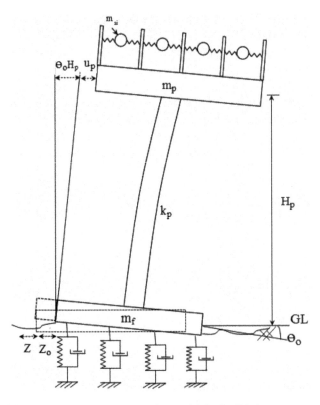

Fig. 6 Lumped mass representative model for flexible-base case

where ω_p is the fixed-base natural frequency and ζ_p is the damping ratio of the aqueduct structure; ω_{si} (with $i = 1, 2, \ldots, n$) are the natural frequencies and ζ_{si} are the damping ratios of water masses; $H_f(\omega)$ is the transfer function of fixed-base aqueduct structure and $h_{si}(\omega)$ (with $i = 1, 2, \ldots, n$) are the transfer functions of fixed-base equivalent convective water masses.

The coefficient of $\ddot{Z}_g(\omega)$ in Eq. (11) is the transfer function say $T_F(\omega)$, for the displacement of fixed-base aqueduct structure $u_p(\omega)$ and horizontal ground acceleration at the base $\ddot{Z}_g(\omega)$.

$$T_F(\omega) = \frac{H_f(\omega)\left(-1 + \frac{1}{m_p}\sum_{i=1}^{n} v_i(\omega)h_{si}(\omega)\right)}{1 + \frac{H_f(\omega)}{m_p}\omega^2 \sum_{i=1}^{n} v_i(\omega)h_{si}(\omega)} \qquad (12)$$

Multiplying each side of Eq. (11) with respective complex conjugates, can be written as the spectral density function (PSDF),

$$u_{pPSD}(\omega) = |T_F(\omega)|^2 \ddot{Z}_{gPSD}(\omega) \qquad (13)$$

The root mean square response $u_{prms}(\omega)$ for the displacement of the aqueduct structure can be calculated as

$$u_{prms}(\omega) = \sqrt{\sum_{i=1}^{N} |T_F(\omega)|^2 \ddot{Z}_{gPSD}(\omega)\Delta\omega} \qquad (14)$$

where $\Delta\omega$ is the appropriate frequency interval and $\omega = \omega_{max}/N$ with N being number of intervals.

Elastic-base case In the flexible-base case (see Fig. 6), when subjected to ground acceleration $\ddot{Z}_g(t)$, the aqueduct experiences forces due to interaction accelerations, $\ddot{Z}_o(t)$ and $\ddot{\theta}_o(t)$, inertial force due to its own mass $m_p\ddot{u}_p(t)$, and forces transferred from the water masses $c_{si}\dot{u}_{si}(t) + k_{si}u_{si}(t)$ (with $i = 1, 2, \ldots, n$). Note that the free-field rocking is neglected. The system equations are as follows for the aqueduct structure and the convective water masses, can be written as follows:

$$m_p\ddot{u}_p(t) + c_p\dot{u}_p(t) + k_pu_p(t) - \sum_{i=1}^{n} k_{si}u_{si}(t) - \sum_{i=1}^{n} c_{si}\dot{u}_{si}(t)$$
$$= -m_p\left(\ddot{Z}_g(t) + \ddot{Z}_o(t) + H_p\ddot{\theta}_o(t)\right) \qquad (15)$$

$$m_{si}\ddot{u}_{si}(t) + c_{si}\dot{u}_{si}(t) + k_{si}u_{si}(t)$$
$$= -m_{si}\left(\ddot{Z}_g(t) + \ddot{Z}_o(t) + H_p\ddot{\theta}_o(t)\right) - m_{si}\ddot{u}_p(t),$$
$$i = 1, 2, \ldots, n \qquad (16)$$

Taking Fourier transform for Eqs. (15) and (16) and on further simplifications, acceleration response $\ddot{u}_p(\omega)$ obtained as follows (see details in Valeti 2013):

$$\ddot{u}_p(\omega) = \omega^2 H_f(\omega)\left(\ddot{Z}_g(\omega) + \ddot{Z}_o(\omega) + H_p\ddot{\theta}_o(\omega)\right.$$
$$\left. -\frac{1}{m_p}\sum_{i=1}^{n} v_i(\omega)u_{si}(\omega)\right) \qquad (17)$$

where $u_p(\omega)$, $u_{si}(\omega)$, $\ddot{Z}_g(\omega)$, $\ddot{\theta}_o(\omega)$ and $\ddot{Z}_o(\omega)$ denote Fourier transforms of $u_p(t)$, $u_{si}(t)$, $\ddot{Z}_g(t)$, $\ddot{\theta}_o(t)$ and $\ddot{Z}_o(\omega)$ respectively. Relationships of foundation accelerations $\ddot{Z}_o(\omega)$, $\ddot{\theta}_o(\omega)$ with hydrodynamic forces $\sum_{i=1}^{n}(k_{si} + i\omega c_{si})u_{si}(\omega)$ and ground acceleration $\ddot{Z}_g(\omega)$ are established using impedance functions, base shear and base moment equations of the aqueduct structure. According to Dey and Gupta (1999), if the foundation is assumed to be massless, then base shear $V_s(\omega)$ and base moment $M_s(\omega)$ can be expressed in terms of foundation displacements relative to the soil medium using the impedance functions.

$$\left\{\begin{array}{c} V_s(\omega) \\ \dfrac{M_s(\omega)}{L} \end{array}\right\} = \left[\begin{array}{cc} S_{xx} & S_{x-ry} \\ S_{ry-x} & S_{ry} \end{array}\right]\left\{\begin{array}{c} Z_o(\omega) \\ L\theta_o(\omega) \end{array}\right\} \qquad (18)$$

Here S_{xx}, S_{ry}, S_{x-ry} and S_{ry-x} are the impedance functions for translation in transverse direction, rotation about the longitudinal direction and coupled translation and rotation of the foundation respectively. As the foundation is assumed to be resting on the surface of the ground, moment due to the reaction from soil on to the side walls of the foundation is neglected. As a result, the coupled impedance functions S_{x-ry} and S_{ry-x} are almost negligible.

Using expressions for base shear and base moment from Eq. (18) in the base shear and base moment equations and substituting $\ddot{u}_p(\omega)$ from Eq. (17) and replacing $Z_o(\omega)$, $\theta_o(\omega)$ by $\frac{\ddot{Z}_o(\omega)}{-\omega^2}$, $\frac{\ddot{\theta}_o(\omega)}{-\omega^2}$ respectively, we obtain equations which contain the terms of only ground translation $(\ddot{Z}_g(\omega))$, relative translational $(\ddot{Z}_o(\omega))$ and rotational $\ddot{\theta}_o(\omega)$ accelerations of foundation with respect to soil and aqueduct-water masses interaction forces.

Solving the resulting simultaneous equations, one can obtain expressions for $\ddot{Z}_o(\omega)$ and $\ddot{\theta}_o(\omega)$ as follows, where Q_1, P_1 represent transfer functions between input free-field ground acceleration $\ddot{Z}_g(\omega)$ and interaction accelerations $\ddot{Z}_o(\omega)$, $\ddot{\theta}_o(\omega)$ respectively. Similarly Q_2, P_2 are the transfer functions between aqueduct-water masses interaction forces $\sum_{i=1}^{n} v_i(\omega)u_{si}(\omega)$ and interaction accelerations $\ddot{Z}_o(\omega)$, $\ddot{\theta}_o(\omega)$ respectively (Ray Chaudhuri and Gupta 2003).

Now, the forces transferred by fixed-base water masses to aqueduct structure, $\sum_{i=1}^{n} v_i(\omega)u_{si}(\omega)$, the transfer function $TF(\omega)$ between relative displacement of aqueduct structure $u_p(\omega)$ and ground acceleration $\ddot{Z}(\omega)$ is obtained as

the response results are hold. After that the transient analyses are performed. These transient analyses are performed using *Newmark*'s integration with $\gamma = 0.5$, $\beta = 0.25$. Unlike frequency domain, damping due to the water is neglected and Rayleigh damping is assumed for the structure. For analysis, *NewtonLineSearch* algorithm is used with a limiting ratio 0.8 and a tolerance of 10^{-18} in the framework of OpenSees (2012).

Results and discussion

The results of the frequency domain and time-history analyses are presented and the observations are discussed as follows.

Frequency domain results

Clough-Penzien power spectral density functions (PSDF) (Villaverde 2009) have been used to represent the input free-field accelerations for soft, medium and firm soil conditions. The parameters for these PSDFs (as given in Table 3) are obtained from Kiureghian and Neuenhofer (1992) except for the medium soil condition for which an interpolation technique is used. The resulting Clough-

$$
TF(\omega) = \frac{H_f(\omega)\left(-\left(1 + Q_1 + H_p P_1\right) + \dfrac{\sum_{i=1}^{n} v_i(\omega)h_{si}(\omega)\left(\frac{-1}{m_p} + Q_2 + H_p P_2\right)\left(1 + Q_1 + H_p P_1\right)}{1 + \left(Q_2 + H_p P_2\right)\sum_{i=1}^{n} v_i(\omega)h_{si}(\omega)}\right)}{1 + H_f(\omega)\dfrac{\sum_{i=1}^{n} v_i(\omega)h_{si}(\omega)\left(\frac{-1}{m_p} + Q_2 + H_p P_2\right)}{1 + \left(Q_2 + H_p P_2\right)\sum_{i=1}^{n} v_i(\omega)h_{si}(\omega)}}
\tag{19}
$$

Time-history analysis

For time-history analysis, the raft foundation resting on the surface of soil, is modeled using *ShallowFoundationGen* command in the framework of OpenSees (2012). To capture the behavior under varying soil conditions, soft, medium and firm states of clayey soils are considered for the analysis. In addition, to study the effect of different base conditions, the foundation is modeled for fixed, linear and nonlinear degrees of flexibility. A vertical factor of safety of $FS_v = 5$ is used for modeling of the foundation.

Transient ground motion analyses are performed for different heights of water for the fixed, linear, and nonlinear base conditions. A gravity analysis is performed and

Penzien PSDFs for all three soil types are shown in Fig. 7. One can observe from this figure that a reduction in amplitude, an increase in spread and a shift towards higher frequencies with an increase in stiffness of the soil.

Fixed-base case Displacement transfer functions relating u_p with ground acceleration for different heights of

Table 3 Clough-Penzien parameters (Kiureghian and Neuenhofer 1992)

Soil type	G_o	ω_g (rad/s)	ω_1 (rad/s)	ζ_g	ζ_1
Soft	0.05	5.0	0.5	0.2	0.6
Medium	0.05	10.0	1.0	0.4	0.6
Firm	0.05	15.0	1.5	0.6	0.6

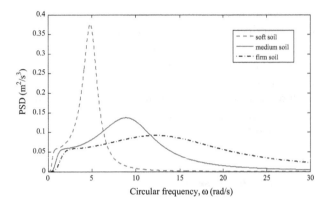

Fig. 7 Clough-Penzien PSDFs for different soil types

water (H) are obtained as shown in Fig. 8. Here, for a given height of water, the first peak from the origin corresponds to the vibration mode of the aqueduct structure whereas the other peaks (on right) represent the sloshing modes. It may be noted that all the sloshing modes (one mode for each of the four channels) are appearing approximately in the same period location. This is because the mass representing the sloshing modes of each channel are equal. A nominal shift towards right corresponding to the first peak (structure) is observed with increasing height of water due to the addition of impulsive water masses. Further, a significant shift towards left is observed for the sloshing modes, indicating its increasing frequency along with the height of water.

Clough-Penzien PSDFs of different soil site conditions are used as the free-field input acceleration PSDFs to obtain the PSDFs of responses of the structure. Figure 9 show the power spectral density (PSD) of u_p with different site conditions. From Fig. 9a–c, the influence of the input PSDFs on u_p can be clearly observed, with a visible peak at period 1.256 s observed in case of soft soil site (at the dominant frequency of the input soft soil site PSDF). Similar peaks are observed at 0.628 s and 0.418 s, respectively, for PSDFs corresponding to medium and firm soil sites. The magnitude of these peaks reduces with an

increase in stiffness of the underlying soil. A similar trend is also observed for base shear PSDFs and not shown here to save space.

RMS values of u_p, base shear, and drift ratio for different heights of water in the aqueduct and different soil sites are obtained. Figure 10 shows the normalized RMS values for the base shear, V_b. Here, the normalization is done with respect to the maximum base shear, i.e., V_b corresponding to the water height $H = 3$ m for firm site. It is observed from this figure that RMS values of displacement for different heights of water do not follow a linear trend. In fact, a small peak is observed for the mid height of water. However, the maximum base shear occurs at full water height. This phenomenon can be explained by looking at the trend of amplitude of peaks at fixed-base natural periods of convective water masses in the transfer function in conjunction with Clough-Penzien PSDF (see Fig. 7). Hence, one can say that the RMS values of displacement are dominated by the response of the convective water masses. A similar trend is also observed for the RMS values of base shear and percentage drift ratio.

Elastic-base case Transfer functions $T_F(\omega)$ and PSDFs are obtained for u_p, base shear, V_b, and drift ratio in all three soil site conditions conditions. Figures 11 and 12 respectively show $T_F(\omega)$ and PSDFs of u_p for different soil site conditions. By comparing Fig. 8 with Fig. 11, one can notice that the peak corresponding to the structure period is almost insignificant in the elastic-base case. This can be clearly observed in Figs. 11d and 12d which, respectively, compare the transfer functions and PSDFs of fixed and elastic base cases (firm soil) for at 1.5 m height of water. This is because of period elongation combined with higher damping due to SSI. But, no such effect of base flexibility is observed on sloshing modes. The significant separation between the structure mode and sloshing modes (sloshing modes are highly flexible compared to the structure mode), changes the structural responses but cannot significantly affect the sloshing mode responses.

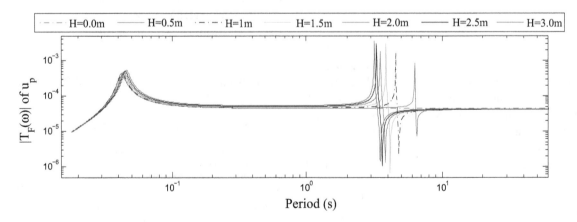

Fig. 8 Transfer functions for u_p evaluated using fixed-base analysis

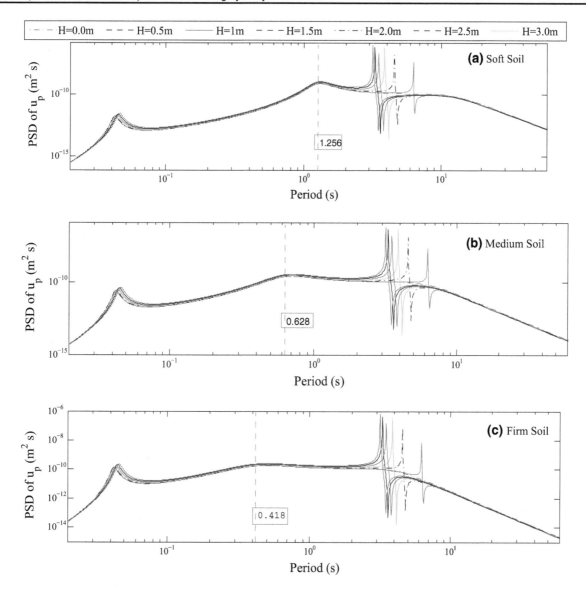

Fig. 9 PSDF of u_p evaluated using fixed-base analysis with underlying soil as: **a** soft, **b** medium and **c** firm

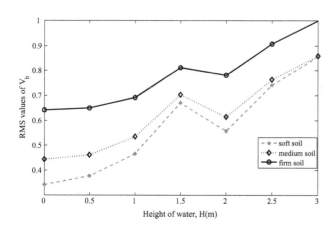

Fig. 10 Normalized RMS values of base shear, V_b evaluated using fixed-base analysis

RMS values for displacement of aqueduct structure (u_p) are calculated from the PSDFs obtained earlier and are normalized for different heights of water and soil conditions. Similarly, the RMS values for the base shear and percentage drift ratio are also obtained. Figure 13 provides the RMS values of V_b with varying water height for different soil site conditions. It is evident that the response of the structure follows the same trend as that of the magnitude of convective water masses of different heights for all soil conditions, as in fixed-base case the input PSDFs have invariant magnitudes for different soil conditions at the natural period of the structure. Also, the base shear increases with an increase in height of water due impulsive masses of water, being highest at the full water level.

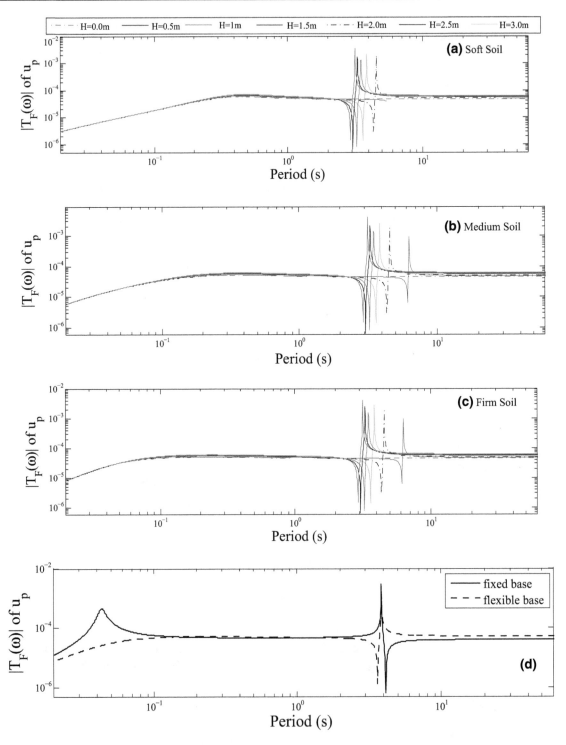

Fig. 11 Transfer functions of u_p (evaluated using flexible-base analysis) for soil conditions: **a** soft, **b** medium, **c** firm and **d** comparison of transfer functions of u_p for water height of 1.5 m and firm soil site (evaluated using fixed base and flexible-base analyses)

Time-history analysis

Sixty records of horizontal acceleration time histories that were generated for the Los Angeles region under FEMA/SAC steel building project (http://www.sacsteel.org/project/) are used as input ground excitations. There are 30 pairs of ground motions with each pair consisting of a fault parallel and a fault normal component of a single ground motion. Out of 30 pairs, there are 10 pairs for each of 3 different hazard levels namely, probability of exceedence of 2 % in 50 years (LA21–LA40), 10 % in 50 years (LA01–LA20) and 50 % in 50 years (LA41–LA60).

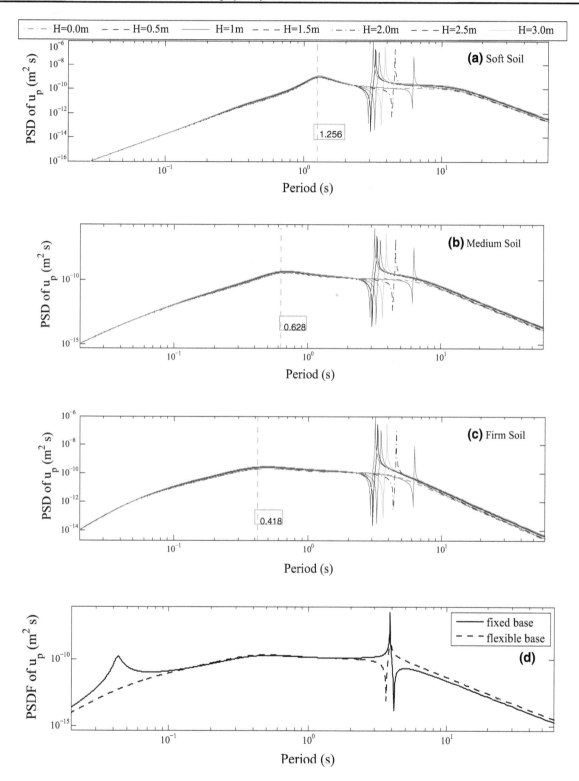

Fig. 12 PSDFs of u_p (evaluated using flexible base analysis) for soil conditions: **a** soft, **b** medium, **c** firm and **d** comparison of PSDFs of u_p for water height of 1.5 m and firm soil site (evaluated using fixed base and flexible-base analyses)

At first, a gravity analysis is performed and then holding the states, time history analysis is performed. Response parameters such as base shear (recorded as the reaction at the base node of the structure) and drift ratio (i.e., relative displacement between the nodes at base and top of the structure normalized by the height) are recorded for comparison. The mean and standard deviation for the peak base shear and peak drift ratio for different heights of water and

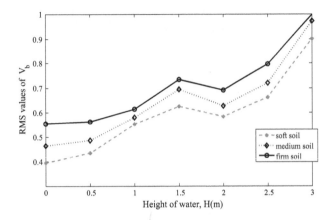

Fig. 13 Normalized RMS values of base shear, V_b (evaluated using flexible-base analysis) for different soil conditions

for each of the three hazard levels of ground motions are evaluated.

Fixed-base case ShallowFoundationGen command with *FootingCondition* 1 is used for the fixed-base case (fixed in all degrees of freedom). The mean (μ) and coefficient of variation (c_v) values of the peak base shear for different heights of water under all three hazard levels of ground motions are shown in Fig. 14. It is found from Fig. 14 that the mean peak base shear values increase with an increase in the height of water and the hazard level of the ground motions. With height of water, the increase in mean values does not follow a linear trend. The local peak at 1.5 m of water (that was earlier observed for the RMS values) is not observed here. This is because the SAC ground motions do not have sufficient energy content to meaningfully excite a sloshing mode. The peak drift ratios (percentage values)

also follow the same trend. The c_v values do not show any trend with water height. One can notice that as the hazard level increases, mean response increases and c_v reduces.

Linear elastic-base case For the time-history analysis of elastic-base case, the *FootingCondition* 3 is used. Gazetas (1991b) static (linear) stiffness values are used at the soil-foundation interface for different degrees of freedom that is sliding, vertical and rotation (Raychowdhury and Hutchinson 2008).

Mean values of peak base shear and drift ratio for different soil conditions and heights of water for different hazard levels are obtained (Fig. 15). In general, for hazard levels of 10 % in 50 years and 50 % in 50 years, the force demands are highest for the firm soil and lowest for soft soils. However, this is not clearly observed for 2 % in 50 years. But, in Fig. 15a one can observe that greater base shear is observed for medium soil than the firm soil at lower heights of water. This can be explained as follows: due to the effect of SSI, flexibility of the structure varies with soil condition leading to period elongation with the increase in softness of the soil. The ground motions in the periods corresponding to medium soil at lower water levels have high energy content. But as the water level increases, the periods shift leading to higher energy content in ground motions at the periods of the firm soil. The displacement demands are lowest for the firm soil and highest for the soft soil as higher flexibility at the base leads to lower base shear (resistance) and higher drift (see Fig. 16). This response is found to increase with the increase in the hazard level of ground motions and height of water.

Nonlinear-base case For nonlinear base condition, *FootingCondition* 5 is used. Nonlinear Winkler springs are

Fig. 14 Peak base shear values for fixed-base case: **a** mean (μ) and **b** coefficient of variation (c_v) for different heights of water, H

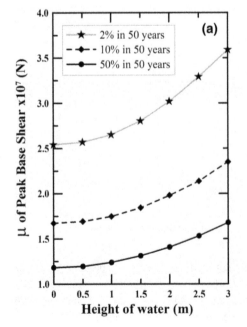

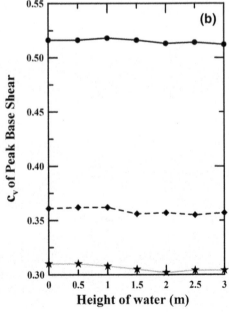

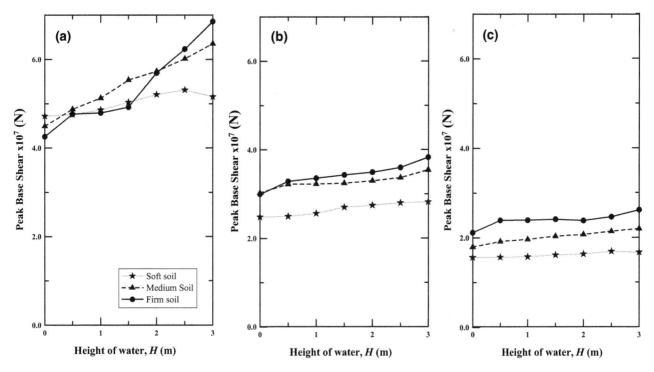

Fig. 15 Mean peak base shear values for (linear) flexible base case at hazard levels: **a** 2 % in 50 years, **b** 10 % in 50 years, and **c** 50 % in 50 years

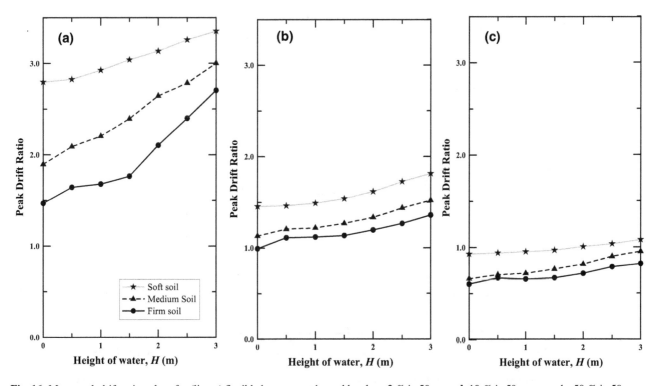

Fig. 16 Mean peak drift ratio values for (linear) flexible base case at hazard levels: **a** 2 % in 50 years, **b** 10 % in 50 years, and **c** 50 % in 50 years

used in the BNWF model representing the soil-foundation system. Mean base shear, mean percentage drift ratio and displacement of the primary structure are recorded in similar fashion to linear base case. Figure 17 show mean peak base shear for different hazard levels, with structure on different soil conditions for nonlinear base case. The

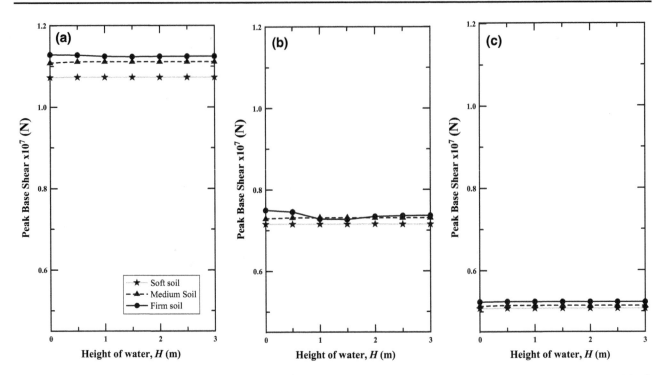

Fig. 17 Mean peak base shear values for (nonlinear) flexible base case at hazard levels: **a** 2 % in 50 years, **b** 10 % in 50 years, and **c** 50 % in 50 years

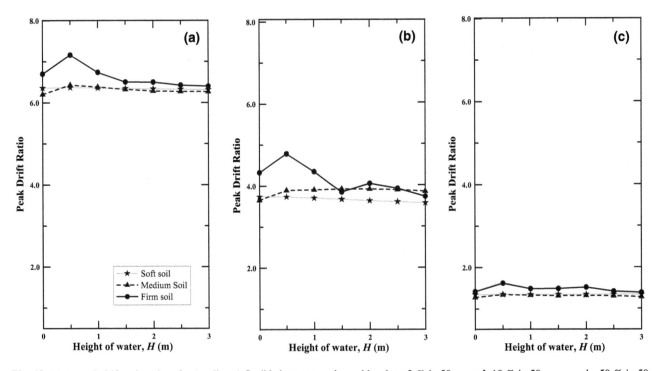

Fig. 18 Mean peak drift ratio values for (nonlinear) flexible base case at hazard levels: **a** 2 % in 50 years, **b** 10 % in 50 years, and **c** 50 % in 50 years

force demands do not increase significantly with increase in height of water as in linear base case (Fig. 17). This is due to the fact that, when the nonlinear Winkler springs reach nonlinear zone (plastic behavior) due to stronger ground motions, their resistance reaches maximum. Similar to linear base case, soft soil shows the lowest base shear,

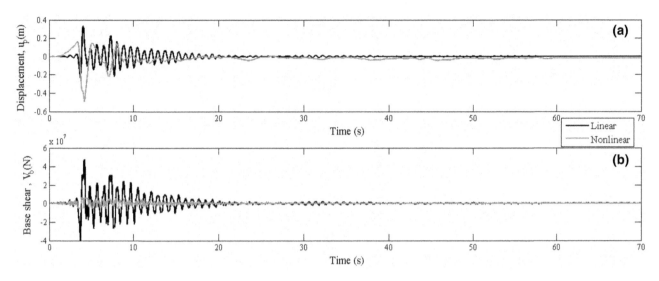

Fig. 19 Time histories of displacement, u_p and base shear V_b for LA-28 with water height of 1.5 m

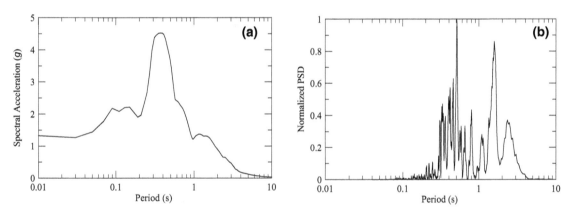

Fig. 20 Characteristics of LA-28 motion: **a** acceleration response spectrum and **b** normalized power spectral density

while firm soil provides the highest base shear. But the trend of drift ratio deviates from that of the linear base case (Fig. 18). This is because of the nonlinear behavior of the soil-foundation interface.

Comparison of linear and nonlinear base cases

Figure 19 shows the comparison of time histories for displacement and base shear demands of linear and nonlinear base cases for LA-28. Figure 20 shows the acceleration response spectrum and normalized power spectral density of LA-28. It can be observed from these figures that this motion contains significant energy around the fundamental period of flexible base aqueduct structure. From Fig. 19, the displacements are observed to be very high for the nonlinear base case when compared to the linear base case. This is because the energy is

dissipated in the nonlinear base case leading to the increased period, displacements and a reduced resistance at the base. A permanent deformation is observed in the displacement time history of nonlinear case. Figure 21 shows the ratio of base shear in nonlinear case to the linear case for different hazard levels and soil conditions. These ratios are found to be less than unity in all the hazard level cases. Similarly, Fig. 22 shows the ratios of drift for nonlinear case to the linear one. All hazard level cases show these ratios greater than unity corroborating the aforementioned explanation. From Figs. 21 and 22, the ratios of both base shear and drift ratio decrease with the increase in height of water. One can thus state that the significance of soil nonlinearity decreases for drift ratio and increases for base shear with an increase in height of water, the highest effect being on firmer soils in both base shear and drift ratio.

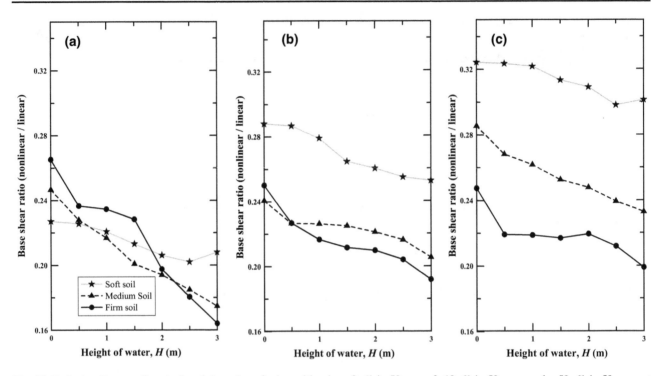

Fig. 21 Ratio (nonlinear to linear) of peak base shear for hazard levels: **a** 2 % in 50 years, **b** 10 % in 50 years and **c** 50 % in 50 years at different water heights

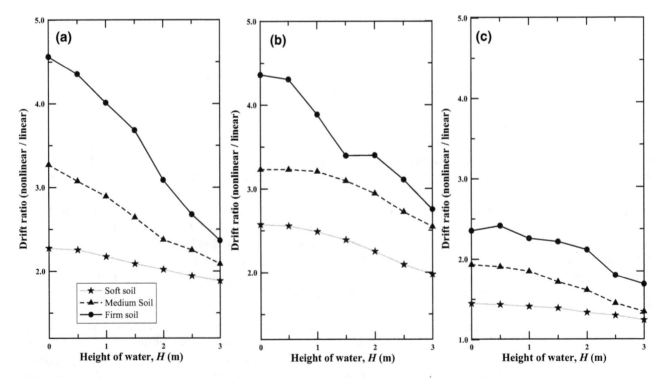

Fig. 22 Ratio (nonlinear to linear) of peak drift ratios for hazard levels: **a** 2 % in 50 years, **b** 10 % in 50 years and **c** 50 % in 50 years at different water heights

Conclusions

In this study, a representative model of an elevated aqueduct is considered with different base fixity conditions. Frequency domain as well as time domain analysis have been performed and demand parameters such as base shear and drift ratio are studied for varying heights of water in channels and different soil conditions at the site. Some of the important findings of this study are as follows:

1. The period of the aqueduct structure increases drastically from 0.025 s in fixed base case to 0.52, 0.59 and 0.743 s for firm, medium and soft soil conditions, respectively. This increase in period from the fixed to the flexible base case can be explained by considering the relative stiffness of the structure (which is very high due to long shear wall) in comparison to the foundation springs. Thus, to capture the dynamic behavior reasonably, it is important to model the base fixity appropriately.

2. With increase in height of water in channels, only a nominal increase in the natural period of the structure is observed. Further, the periods of convective masses, which are high enough compared to the period of the structure, are not affected significantly due to change in base conditions.

3. In case of fixed-base analysis, the RMS values of response for different water depths are observed to follow the same trend as that of the amplitude of its transfer function at the natural frequencies of the convective water masses. This indicates high influence of the convective mass on the response of the structure for frequency domain analysis. This influence of the convective mass is not however observed for the time history analysis of the fixed base case. This is because in time history analysis, the ground motions do not have significant energy to excite the convective modes.

4. For frequency domain analysis of elastic base case, the base shear demand is found to be higher for firmer soil conditions. Further, response increases with the increase in height of water and decrease in soil flexibility. A similar trend is also observed from the time history analysis.

5. Introduction of nonlinearity at soil-foundation interface leads to increase in drift ratio and decrease in base shear compared to the linear base case. Negligible effect on the response due to change in the height of water is found for the nonlinear case due to mobilization of soil at the base.

6. The decrease in base shear is higher in firmer soils from linear to nonlinear base case for ground motions with 50 % in 50 years hazard level. This trend is however not significant for ground motions with 2 % in 50 years hazard levels due significant nonlinearity at the soil-foundation interface for such a high hazard level.

7. From time history analysis, it is observed that for all hazard levels, base shear is lower for the nonlinear base case than the elastic base case. The trend is opposite for drift ratio, i.e., drift ratio is higher for nonlinear base case in comparison to the linear base case. Also, this ratio of nonlinear to linear responses (of base shear and drift) reduces with an increase in height of water in the channels. This is because, for varying water levels, these responses are least affected in nonlinear base case and get altered only in case of elastic base. In general, for a given hazard level (say 10 % in 50 years), the base shear demand of empty aqueduct is maximum for elastic base case, followed by fixed base and nonlinear base cases.

The results of this study are limited to the parameter space considered, ground motions and PSDFs used, and modeling assumptions. Further, this study does not consider the hydrodynamic effects that may be present in case the piers are submerged in water. The findings of this study may however provide a better understanding of seismic behavior of an elevated aqueduct under various modeling assumptions and input excitations.

References

Boulanger R, Curras C, Kutter B, Wilson D, Abghari A (1999) Seismic soil-pile-structure interaction experiments and analyses. ASCE J Geotech Geoenvironmental Eng 125(9):750–759

Bowles J (1988) Foundation analysis and design. McGraw-Hill Publishing Company, USA

Dey A, Gupta VK (1999) Stochastic seismic response of multiply-supported secondary systems in flexible-base structures. Earthq Eng Struct Dynam 28(4):351–369

Gazetas G (1991a) Formulas and charts for impedances of surface and embedded foundations. ASCE J Geotech Eng 117(9):1363–1381

Gazetas G (1991b) Foundation vibrations. In: Foundation Engineering Handbook. Springer

Gutierrez JA, Chopra AK (1978) Evaluation of methods for earthquake analysis of soil-structure interaction. In: Proceedings of the Sixth World Conference on Earthquake Engineering, vol 6. Berkely

Harden C, Hutchinson T, Martin GR, Kutter BL (2005) Numerical modeling of the nonlinear cyclic response of shallow foundations. In: Tech. rep., Pacific Earthquake Engineering Research Center, College of Engineering. University of California, Berkeley

Housner GW (1954) Earthquake pressures on fluid containers. Tech. rep. California Institute of Technology, Pasadena

Idelsohn S, Onate E, Del Pin F, Calvo N (2006) Fluid-structure interaction using the particle finite element method. Comput Methods Appl Mech Eng 195:2100–2123

Kiureghian AD, Neuenhofer A (1992) Response spectrum method for multi-support seismic excitations. Earthq Eng Struct Dynam 21(8):713–740

Liu WK (1981) Finite element procedures for fluid-structure interactions and application to liquid storage tanks. Nucl Eng Des 65(2):221–238

OpenSees (2012) Open system for earthquake engineering simulation: OpenSees. Pacific Earthquake Engineering Research Center (PEER), Richmond

Ramaswamy B, Kawahara M (1987) Arbitrary Lagrangian-Eulerianc finite element method for unsteady, convective, incompressible viscous free surface fluid flow. Int J Numer Meth Fluids 7(10):1053–1075

Ray Chaudhuri S, Gupta VK (2003) Mode acceleration approach for generation of floor spectra including soil-structure interaction. Int J Earthq Technol 40:99–115

Raychowdhury P (2011) Seismic response of low-rise steel moment-resisting frame (SMRF) buildings incorporating nonlinear soil-structure interaction (SSI). Eng Struct 33:958–967

Raychowdhury P, Hutchinson TC (2009) Performance evaluation of a nonlinear winkler-based shallow foundation model using centrifuge test results. Earthq Eng Struct Dynam 38:679–698

Raychowdhury P, Singh P (2012) Effect of nonlinear soil-structure interaction on seismic response of low-rise SMRF buildings. Earthq Eng Eng Vib 11(4):541–551

Raychowdhury P, Hutchinson TC (2008) ShallowFoundationGen Opensees Documentation

Valeti B (2013) Seismic analysis of a river aqueduct considering soil-structure interaction and hydrodynamic effects. Master's thesis, Indian Institute of Technology Kanpur, Kanpur, UP, 208016, India

Villaverde R (2009) Fundamental Concepts of Earthquake Engineering. Taylor and Francis Group, UK

Westergaard H (1933) Water pressures on dams during earthquakes. Trans Am Soc Civ Eng 98(2):418–433

Wong HL, Luco JE (1985) Tables of impedance functions for square foundations on layered media. J Soil Dyn Earthq Eng 4(2):64–81

9

Vibration control of bridge subjected to multi-axle vehicle using multiple tuned mass friction dampers

Alka Y. Pisal[1] · R. S. Jangid[2]

Abstract The effectiveness of tuned mass friction damper (TMFD) in reducing undesirable resonant response of the bridge subjected to multi-axle vehicular load is investigated. A Taiwan high-speed railway (THSR) bridge subjected to Japanese SKS (Salkesa) train load is considered. The bridge is idealized as a simply supported Euler–Bernoulli beam with uniform properties throughout the length of the bridge, and the train's vehicular load is modeled as a series of moving forces. Simplified model of vehicle, bridge and TMFD system has been considered to derive coupled differential equations of motion which is solved numerically using the Newmark's linear acceleration method. The critical train velocities at which the bridge undergoes resonant vibration are investigated. Response of the bridge is studied for three different arrangements of TMFD systems, namely, TMFD attached at mid-span of the bridge, multiple tuned mass friction dampers (MTMFD) system concentrated at mid-span of the bridge and MTMFD system with distributed TMFD units along the length of the bridge. The optimum parameters of each TMFD system are found out. It has been demonstrated that an optimized MTMFD system concentrated at mid-span of the bridge is more effective than an optimized TMFD at the same place with the same total mass and an optimized MTMFD system having TMFD units distributed along the length of the bridge. However, the distributed MTMFD system is more effective than an optimized TMFD system, provided that TMFD units of MTMFD system are distributed within certain limiting interval and the frequency of TMFD units is appropriately distributed.

Keywords Bridge · Multi-axle vehicle · Critical velocity · Resonant response · TMFD · Concentrated MTMFD · Distributed MTMFD

Introduction

Transportation infrastructure is one of the significant factor which reflects the development of a nation's economy. Due to the paucity of land and increased traffic in urban areas, the bridges have become inevitable part of the transportation facilities, such as highways and railways. In recent years, with the rapid advances in the area of high-performance materials, design technologies and construction techniques, the architecture of bridges has reached unexpected limits. At the same course of time, day by day, the bridges are becoming more slender and lighter and hence more prone to vibrations due to heavy vehicles and high-speed trains passing over it. Thus, the vibration of a bridge due to the passage of vehicles is an important aspect in bridge design. A multi-axle vehicle moving over the bridge can be modeled as planar moving forces, inducing periodic excitations to the bridge. Vibrations induced by moving vehicles become excessive when the vehicle velocities reach resonant or critical values. This may seriously affect the long-term safety, serviceability of the bridges and comfort of the passengers. In addition, it may endanger the safety of supporting structure. Hence, it is very necessary to control these undesirable excessive vibrations of bridge under train loads.

Literature review shows that the dynamic behavior of the bridges has been significantly impacted due to periodic

⊠ Alka Y. Pisal
  alka.y.pisal@gmail.com

[1] Department of Civil Engineering, D. Y. Patil College of Engineering, Akurdi, Pune 411044, India

[2] Department of Civil Engineering, Indian Institute of Technology Bombay, Powai, Mumbai 400076, India

moving loads from trains; the same has been investigated by many researchers in the past few years.

Kwon et al. (1998) investigated the control efficiency of single-tuned mass damper (STMD) system attached with the bridges. He idealized the bridge as a simply supported Euler–Bernoulli beam traversed by vehicles which were modeled as moving masses and shown that TMD can effectively reduce response of the bridge. Yang et al. (1997) investigated the vibration of simple beam excited under high-speed moving trains and proposed a span to car length ratio so that no resonance occurs in beam. Chen and Lin (2000) investigated the dynamic response of elevated high-speed railway. Cheng et al. (2001) proposed a bridge-track-vehicle element for investigating interactions between moving train, railway track and bridge. Ju and Lin (2003) investigated the resonant characteristics of three-dimensional (3D) multi-span bridges subjected to high-speed trains. Wang et al. (2003) carried out the optimization of STMD attached at the mid-span of Taiwan high-speed railway (THSR) bridges subjected to French train à grande vitesse (TGV), German intercity express (ICE) and Japanese SKS trains. Nasiff and Liu (2004) presented a 3D dynamic model for the bridge-road-vehicle interaction system. Jianzhong et al. (2005) has performed a parametric study on the optimization of multiple tuned mass dampers (MTMD) and demonstrated its efficiency in reducing displacement and acceleration responses of simply supported bridge subjected to high-speed trains. Lin et al. (2005) proved that MTMD is more effective and reliable than STMD in reducing train-induced dynamic responses of simply supported bridges during resonant speeds. Shi and Cai (2008) performed numerical analysis to study the vehicle-induced bridge vibration response using a TMD, considering the road surface conditions. Li et al. (2010) developed a numerical method to analyze coupled railway vehicle-bridge systems of non-linear features. Moghaddas et al. (2012) studied the dynamic behavior of bridge-vehicle system attached with TMD, using the finite-element method. It was shown that by attaching an optimized TMD to a bridge, a significantly faster response reduction can be achieved. Antolin et al. (2013) considered non-linear wheel-rail contact forces model to analyze the dynamic interaction between high-speed trains and the bridges. Wang et al. (2013) investigated the effectiveness of visco-elastic damper (VED) to mitigate the multiple resonant responses of a moving train running on two-span continuous bridges. It is found that with the installation of VED at midpoint of each span, the maximum acceleration response of the bridge can be suppressed noticeably at resonant speeds.

It is evident from above studies that dynamic forces in the form of periodic moving loads from train may result in undesirable resonant responses of the bridge. These undesirable responses can be controlled using the TMD, MTMD

and VED. However, TMD and MTMD are having disadvantages of being sensitive to the fluctuation in frequency spacing and tuning frequency with respect to the controlling (fundamental) frequency of the bridge. This may deteriorate the performance of TMD and MTMD. Until the date, the resonant response control of bridge is studied using TMD, MTMD and VED, but the use of friction dampers (FD) and tuned mass friction damper (TMFD) has not been explored for controlling the response of the bridges. Pisal and Jangid (2015) demonstrated that MTMFD can effectively reduce the seismic response of multi-storey structures.

In the present study, the performance of tuned mass friction damper (TMFD) in controlling undesirable resonant response of the bridge is investigated. To suppress the undesirable excessive responses of the bridge, different TMFD systems are employed. The specific objectives of the study are summarized as:

1. To formulate the equation of motion for the response of the bridge with different TMFD systems under periodic train loads and develop its solution procedure.
2. To investigate the influence of important parameters, such as mass ratio, tuning frequency ratio, frequency spacing, damper slip force and number of TMFD units in MTMFD on the performance of the MTMFD.
3. To obtain the optimum values of influencing parameters for TMFD and MTMFD systems, which may find application in the effective design of MTMFD for the bridges.
4. To investigate the performance of different TMFD systems, i.e., TMFD attached at the mid-span of the THSR bridge, MTMFD attached at the mid-span (concentrated) of the THSR bridge, and MTMFD distributed along the length of the THSR bridge subjected to Japanese SKS trains.

Modeling of vehicle, bridge and MTMFD

A train or any other vehicle which has multi-axle system excites a bridge when it passes over it. In contrast to wind or earthquake load, the position of vehicular load over the bridge changes at each and every second. Furthermore, due to the interaction between the vehicle and the bridge, the magnitude of vehicular load depends on the response of the bridge. Thus, it is difficult to establish the correlation between governing parameters and response of the bridge. Furthermore, time variant properties of the configured vehicle loading, bridge and TMFD system can make the modeling complex. To get an efficient and effective control performance of the TMFD system, it is necessary to select governing parameters appropriately. In the current study, simplified modeling of the vehicle and the bridge is considered to determine governing parameters. Once governing parameters are identified, model needs to be refined for doing further research. Finally, based on the simplified model of the vehicle, bridge and TMFD, coupled differential equations of motion is derived, the same has been solved numerically using the Newmark's linear acceleration method.

The bridge is idealized as a simply supported Euler–Bernoulli beam with uniform properties throughout the length of the bridge. Track irregularity is neglected in the conceptualization of the model. The vibration of bridge is considered only in the vertical translational direction. The bridge is a continuous system which has infinite degrees-of-freedom (DOF), but only the first few modes of the bridge contribute significantly to the total dynamic response. Among all modes, the fundamental mode is dominating mode, especially for the displacement response of a simply supported bridge. Thus, for the present study, fundamental mode has been considered while finding out peak mid-span response of the bridge.

There are three basic modeling procedures of vehicles, namely, moving force, moving mass and moving suspension mass, as shown in Fig. 1 (Wang et al. 2003). In this

study, the real-life multi-axle vehicle is modeled as a set of moving forces having equal axle spacing, moving along the centre line of the bridge. The axle spacing is an important parameter in the determination of critical resonating velocities of the vehicle which causes excessive vibrations in the bridge.

Arrangement considered for the present study consists of the bridge as primary system which is attached with TMFD and MTMFD with different dynamic characteristics. Here, the bridge is modeled as a simply supported Euler–Bernoulli beam.

Since, the maximum response of a simply supported bridge occurs at its mid-span, a TMFD system is installed at the mid-span of the bridge. In the case of MTMFD system, all the TMFD units can be concentrated at the mid-span or can be distributed along the length of the bridge. Figure 2 shows the simplified model of the bridge with MTMFD attached at equal intervals under the bridge and subjected to a train-induced excitation. The multi-axle train load is modeled as moving force with equal axle spacing. Assumptions/considerations are proposed for the present study in the following sub points:

1. Stiffness of each TMFD unit is the same.
2. Normalized slip force value of each TMFD unit is the same.
3. The mass of each TMFD unit is varying. Thus, the natural frequency of each TMFD unit is adjusted to the required value by varying the mass.
4. The natural frequencies of the TMFD units in an MTMFD system are uniformly distributed around their average natural frequency. It is to be noted that the TMFD units of a MTMFD system with identical dynamic characteristics are equivalent to a TMFD, in which the natural frequency of the individual TMFD units in a MTMFD is the same as that of the equivalent TMFD.
5. For a simply supported bridge, dominating mode is the fundamental mode; hence, both TMFD as well as MTMFD systems are tuned to the fundamental natural frequency of the bridge.

Let ω_T be the average frequency of all MTMFD and it can be expressed as:

$$\omega_T = \sum_{j=1}^{r} \frac{\omega_{dj}}{r}, \tag{1}$$

where r is the total number of TMFD units in MTMFD, and ω_{dj} is the natural frequency of the jth TMFD and it can be expressed as:

$$\omega_{dj} = \omega_T \left[1 + \left(j - \frac{r+1}{2} \right) \right] \frac{\beta}{r-1}, \tag{2}$$

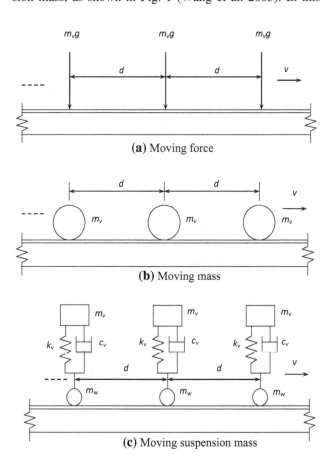

$m_v g$ $m_v g$ $m_v g$

d d v

(a) Moving force

d d v

m_v m_v m_v

(b) Moving mass

m_v m_v m_v

k_v c_v k_v c_v k_v c_v

d d v

m_w m_w m_w

(c) Moving suspension mass

Fig. 1 Schematic modeling of vehicles (Wang et al. 2003)

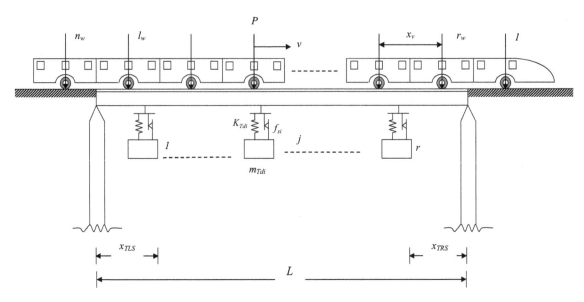

Fig. 2 Simplified model of bridge with MTMFD under multi-axle vehicle

where β is the dimensionless frequency spacing of the MTMFD, given as:

$$\beta = \frac{\omega_r - \omega_1}{\omega_T}. \tag{3}$$

If k_{dj} is the constant stiffness of each TMFD unit, then the mass of the jth TMFD unit is expressed as:

$$m_{dj} = \frac{k_{dj}}{\omega_{dj}^2}. \tag{4}$$

The mass ratio, which is defined as the ratio of total MTMFD mass to total mass of the bridge, is expressed as:

$$\mu = \frac{\sum_{j=1}^{r} m_{dj}}{mL}, \tag{5}$$

where m is the mass per-unit length of the bridge, and L is the total length of bridge.

The ratio of average frequency of the MTMFD to the fundamental frequency of bridge is defined as tuning frequency ratio and is expressed as

$$f = \frac{\omega_T}{\omega_1}, \tag{6}$$

where ω_1 is the fundamental frequency of the bridge. It is to be noted that as the stiffness and the normalized damper forces of all the TMFD are constant and only mass ratio is varying, the friction force adds up. Thus, the non-dimensional frequency spacing β controls the distribution of the frequency of TMFD units.

Governing equations of motion and solution procedure

Arrangement considered for the present study consists of the bridge as primary system which is subjected to a train-induced excitation. The train load is modeled as multi-axle moving force having equal axle spacing, as shown in Fig. 2. It is assumed that the train is compiled by total n_w numbers of axles with equal axle spacing of x_v and the numbering of axles is done in an ascending order from right to left. The rightmost and the leftmost axles over the bridge are numbered as r_w and l_w, respectively; each axle is transmitting a concentrated axle load of intensity P. The MTMFD is arranged under the bridge at equal interval of x_T, with the leftmost TMFD at a distance of x_{TLS} from the left support and the rightmost TMFD at a distance of x_{TRS} from the right support. Each TMFD is a single-degree-of-freedom (SDOF) system pertaining to vibration along the vertical translational direction. The governing differential equation of the vertical motion of the bridge is:

$$\frac{\partial^2}{\partial x^2}\left[E_m I_m(x)\frac{\partial^2 z_b(x,t)}{\partial x^2}\right] + m(x)\frac{\partial^2 z_b(x,t)}{\partial t^2} + c(x)\frac{\partial z_b(x,t)}{\partial t}$$
$$= P_V(x,t) + P_T(x,t), \tag{7}$$

where E_m is the modulus of elasticity of the bridge material; $I_m(x)$ is the moment of inertia of uniform cross section of the bridge; c is the damping co-efficient of the bridge in the vertical direction; $P_V(x, t)$ is interacting force between the vehicle and the bridge; $P_T(x, t)$ is interacting force between the TMFD and the bridge and $Z_b(x, t)$ is vertical displacement of the bridge at a distance x and at time t.

$$P_V(x,t) = \sum_{i=r_w}^{l_w} P\delta\big(x - (vt - (i - 1)x_V)\big), \qquad (8)$$

$$P_T(x,t) = \sum_{j=1}^{r} \left[\begin{array}{l} K_{dj}\big(Z_{dj} - Z_b((x_{TLS} + (j-1)x_T),t)\big) \\ + f_{sj}\,\mathrm{sgn}\left(\dot{Z}_{dj} - \dfrac{\delta Z_b\big(Z_{dj} - Z_b((x_{TLS} + (j-1)x_T),t)\big)}{\partial t}\right) \end{array} \right]$$
$$\times \delta\left(x - (x_{TLS} + (j-1)x_T)\right), \qquad (9)$$

where P is the vehicular load, $\delta()$ is the Dirac delta function, v is the velocity of the vehicle, x_v is the axle spacing of vehicle, sgn denotes the signum function, Z_{dj} is the vertical displacement of the jth TMFD and $\dot{Z}_{dj}$ is the velocity of the jth TMFD.

The equation of motion of the jth TMFD can be expressed as:

$$m_{dj}\ddot{Z}_{dj} + K_{dj}\left(Z_{dj} - Z_b((x_{TLS} + (j-1)x_T),t)\right)$$
$$= -f_{sj}\,\mathrm{sgn}\left(\dot{Z}_{dj} - \dot{Z}_b((x_{TLS} + (j-1)x_T),t)\right), \qquad (10)$$

where $\dot{Z}_b$ is the vertical velocity of the bridge, and $\ddot{Z}_{dj}$ is the vertical acceleration of the jth TMFD.

Modal analysis has been carried out to separate governing parameters and solve Eqs. (7) and (10) analytically. The vertical displacement of the bridge can be expressed as the product of mode shape function, $\varphi_n(x)$ and modal response function, $q_n(t)$ involving only the spatial and temporal co-ordinates

$$Z_b(x,t) = \sum_{i=1}^{n} \varphi_n(x)q_n(t), \qquad (11)$$

where n is the number of modes to be considered for bridge. Substituting Eq. (11) in Eq. (7) results in Eq. (12)

$$m\sum_{i=1}^{n} \varphi_n(x)\ddot{q}_n(t) + c\sum_{i=1}^{n} \varphi_n(x)\dot{q}_n(t)$$
$$+ \sum_{i=1}^{n} \left[E_m I_m \varphi_n''(x)\right]'' q_n(t) = P_V(t) + P_T(t). \qquad (12)$$

Multiplying each term of Eq. (12) by the mode shape function and integrating over the length of the bridge, we can obtain the following equation, after applying orthogonality principle of mode shape function for the nth mode of vibration of the bridge

$$M_n\ddot{q}_n(t) + C_n\dot{q}_n(t) + K_n q_n(t) = P_{nV}(x,t) + P_{nT}(x,t), \qquad (13)$$

where M_n, K_n and C_n are representing the modal mass, modal stiffness and modal damping of the nth mode, respectively; $q_n(t)$, $\dot{q}_n(t)$ and $\ddot{q}_n(t)$ represent the modal displacement, modal velocity and modal acceleration of the bridge in the nth mode of vibration, respectively. The modal interacting forces for the nth mode become:

$$P_{nV}(x,t) = \int_0^L \sum_{i=r_w}^{l_w} P\varphi_n(x)\,\delta\left(x - (vt - (i-1)x_V)\right)$$
$$= \sum_{i=r_w}^{l_w} P\varphi_n(vt - (i-1)x_V), \qquad (14)$$

$$P_{nT}(x,t) = \sum_{j=1}^{r} \left[\begin{array}{l} K_{dj}\left(Z_{dj} - \displaystyle\sum_{i=1}^{n} \varphi_i(x_{TLS} + (j-1)x_T)\,q_i(t)\right) \\ + f_{sj}\,\mathrm{sgn}\left(\dot{Z}_{dj} - \displaystyle\sum_{i=1}^{n} \varphi_i(x_{TLS} + (j-1)x_T)\,\dot{q}_i(t)\right) \end{array} \right]$$
$$\varphi_n\left(x_{TLS} + (j-1)x_T\right). \qquad (15)$$

The natural frequencies and the mode shape functions for the nth mode of a simply supported bridge are given as:

$$\omega_n = \frac{n^2\pi^2}{L^2}\sqrt{\frac{E_m I_m}{m}}, \qquad (16)$$

$$\varphi_n(x) = \sin\left(\frac{n\pi x}{L}\right). \qquad (17)$$

The modal mass is the same for all the modes and is equal to half total mass of the bridge. Modify Eq. (13) by substituting $M_n = mL/2$ and replacing suffix n by 1 to obtain the equation of motion for the fundamental mode of vibration.

$$\frac{mL}{2}\ddot{q}_1 + 2\left(\frac{mL}{2}\right)\xi_1\omega_1\dot{q}_1 + \frac{mL}{2}\omega_1^2 q_1 = \sum_{i=r_w}^{l_w} P\varphi_1(vt - (i-1)x_V)$$
$$+ \sum_{j=1}^{r} \left[\begin{array}{l} K_{dj}\big(Z_{dj} - \varphi_1(x_{TLS} + (j-1)x_T)\,q_1\big) \\ + f_{sj}\,\mathrm{sgn}\big(\dot{Z}_{dj} - \varphi_1(x_{TLS} + (j-1)x_T)\,\dot{q}_1(t)\big) \end{array} \right]$$
$$\times \varphi_1\left(x_{TLS} + (j-1)x_T\right). \qquad (18)$$

Similarly, the equation of motion for TMFD can be written as Eq. (19) by modifying Eq. (10)

$$m_{dj}\ddot{Z}_{dj} + K_{dj}\big(Z_{dj} - \varphi_1(x_{TLS} + (j-1)x_T)\,q_1\big)$$
$$= -f_{sj}\,\mathrm{sgn}\big(\dot{Z}_{dj} - \varphi_1(x_{TLS} + (j-1)x_T)\,\dot{q}_1\big). \qquad (19)$$

Equation (20) represents the coupled equation of motion in matrix form for the bridge equipped with TMFD. This equation is obtained by combining Eqs. (18) and (19)

$$[M_1]\{\ddot{y}_1\} + [C_1]\{\dot{y}_1\} + [K_1]\{y_1\} = [E]\{P_{1V}\} + [B]\{F_s\}, \qquad (20)$$

$$\{y_1\} = \begin{Bmatrix} q_1 \\ Z_{d1} \\ Z_{d2} \\ \vdots \\ Z_{dr} \end{Bmatrix}, \tag{21}$$

where q_1 is the first modal displacement vector of the bridge, and Z_{dj} ($j = 1$ to r) is the displacement vector of TMFD units of MTMFD; $[M_1]$, $[C_1]$ and $[K_1]$ denote the mass, damping and stiffness matrix of the configured system of order $(r + 1) \times (r + 1)$, respectively, considered for the study for the fundamental mode of vibration of bridge; $[E]$ and $[B]$ are placement matrices for the train-induced excitation force and friction force, respectively; y, $\dot{y}$ and y are the vertical displacement, velocity and acceleration vector of configured system, respectively; P_{1v} is the interacting force vector between the vehicle and the bridge in fundamental mode and F_s denotes the vector of friction force provided by the TMFD. These matrices can be shown as:

$$[M_1] = \text{diag} \left[\frac{mL}{2} \; m_1 \; m_2 \cdots m_r \right], \tag{22}$$

$$[C_1] = \begin{bmatrix} mL\xi_1\omega_1 & 0 & 0 & \cdots & 0 \\ 0 & 0 & 0 & \cdots & 0 \\ 0 & 0 & 0 & \cdots & 0 \\ \vdots & \vdots & \vdots & \ddots & \vdots \\ 0 & 0 & 0 & \cdots & 0 \end{bmatrix}, \tag{23}$$

$$F_{sj} = f_{sj} Z, \tag{27}$$

where f_{sj} is the limiting friction force or slip force of the jth TMFD, and Z is the non-dimensional hysteretic component, which satisfies the following first-order non-linear differential equation:

$$q \frac{dZ}{dt} = A \left(\dot{Z}_{dj} - \dot{q}_1 \right) - \beta \left| \left(\dot{Z}_{dj} - \dot{q}_1 \right) \right| Z \, |Z|^{n-1} - \tau \left(\dot{Z}_{dj} - \dot{q}_1 \right) |Z|^n, \tag{28}$$

where q represents the yield displacement of frictional force loop, and A, β, τ and n are non-dimensional parameters of the hysteretic loop which controls the shape of the loop. These parameters are selected in such a way that it provides typical Coulomb-friction damping. The recommended values of these parameters are taken as $q = 0.0001$ m, $A = 1$, $\beta = 0.5$, $\tau = 0.05$, and $n = 2$ (Bhaskararao and Jangid 2006). The hysteretic displacement component, Z, is bounded by peak values of ± 1 to account for the conditions of sliding and non-sliding phases. The limiting friction force or slip force of the jth friction damper is expressed in the normalized form by R_{fj}, which can be expressed as:

$$R_{fj} = \frac{f_{sj}}{m_{dj} \times g}, \tag{29}$$

$$[K_1] = \begin{bmatrix} \frac{mL}{2}\omega_1^2 + \sum\limits_{j=1}^{n} K_{dj}\varphi_1^2(x_{\text{TLS}} + (j-1)x_{\text{T}}) & -K_{d1}\varphi_1(x_{\text{TLS}}) & -K_{d2}\varphi_1(x_{\text{TLS}} + x_{\text{T}}) & \cdots & -K_{dr}\varphi_1(x_{\text{TLS}} + (r-1)x_{\text{T}}) \\ -K_{d1}\varphi_1(x_{\text{TLS}}) & K_{d1}\varphi_1(x_{\text{TLS}}) & 0 & \cdots & 0 \\ -K_{d2}\varphi_1(x_{\text{TLS}} + x_{\text{T}}) & 0 & K_{d2}\varphi_1(x_{\text{TLS}} + x_{\text{T}}) & \cdots & 0 \\ \vdots & \vdots & \vdots & \cdots & 0 \\ -K_{dr}\varphi_1(x_{\text{TLS}} + (r-1)x_{\text{T}}) & 0 & 0 & 0 & K_{dr}\varphi_1(x_{\text{TLS}} + (r-1)x_{\text{T}}) \end{bmatrix}, \tag{24}$$

$$F_s = \left[\sum_{j=1}^{r} F_{sj} \quad -F_{s1} \quad -F_{s2} \quad \cdots \quad -F_{sr} \right], \tag{25}$$

where the friction force of the jth damper is given as:

$$F_{sj} = f_{sj} \, \text{sgn} \left(\dot{Z}_{dj} - \varphi_1(x_{\text{TLS}} + (j-1)x_{\text{T}}) \dot{q}_1 \right), \tag{26}$$

where $\dot{Z}_{dj}$ shows the velocity of the jth TMFD, and $\dot{q}_1$ shows the first modal velocity of the bridge. Furthermore, the damper forces are calculated using the hysteretic model proposed by Constantinou et al. (1990), using the Wen's equation (Wen 1976), which is expressed as:

where g represents acceleration due to gravity.

The coupled differential equations are solved using the Newmark's linear acceleration method (Chopra 2003).

Critical velocities of train

Dynamic response of the bridge becomes excessive under resonating conditions, i.e., when the vehicle velocities are critical. The critical velocities of the vehicle depends on the fundamental frequency of the bridge and axle spacing of the vehicle which can be expressed as (Wang et al. 2003):

$$v_c = \frac{\omega_1 x_v}{2l\,\pi},$$ (30)

where ω_1 is the fundamental frequency of the bridge, x_v is the vehicle's axle spacing and $l = 1, 2, 3, \ldots$ etc. At these critical velocities of the vehicle, the dynamic response of the bridge becomes excessive, generally when $l = 1$. Thus, the resonant responses occurring due to the first critical velocity v_c for $i = 1$ are of major concern.

Numerical study

The THSR bridge that is considered for the numerical study and properties of this bridge is listed in Table 1 (Wang et al. 2003). The bridge is subjected to Japanese SKS train, modeled as a series of moving planar forces with the same magnitude of axle load and equal axle spacing. The properties of this Japanese SKS train have been presented in Table 2. The bridge undergoes resonant vibration whenever the velocity of the train reaches its critical value. The response quantity of interest for the study is the mid-span vertical displacement of the bridge.

For numerical study, TMFD and MTMFD systems are installed under the bridge. Maximum displacement response of the simply supported bridge occurs at mid-span; hence for the proposed TMFD system, the damper is placed at the mid-span of the bridge. In the case of MTMFD systems, all the TMFD units can be concentrated at the mid-span or can be distributed at an equal interval along the length of the bridge. Furthermore, for distributed MTMFD system, the TMFD units are distributed at an interval of 2 and 5 m, respectively. Time interval, $\Delta t = 0.001$ has been considered for numerical solution. The performance of the bridge installed with TMFD system has been compared with the uncontrolled response of the bridge.

Uncontrolled response of the bridge under train load

The generalized responses of the first three modes of the THSR bridge without the installation of TMFD under the effect of Japanese SKS train induced vibration is studied. It is seen that both dynamic displacement as well as dynamic acceleration responses are dominated by fundamental mode and the contribution of higher modes can be neglected to approach at the feasible solution for response. The contribution of the first three modes towards vertical displacement of THSR bridge subjected to Japanese SKS train is 5.249, 0.205 and 0.026 mm, respectively, and that towards acceleration response is 3.028, 1.301 and 0.122 m/s², respectively. Therefore, it is apparent that only fundamental mode requires to be considered especially for the displacement response in practice. As excessive dynamic displacement affects the long-term safety, serviceability of the bridge and comfort of the passenger, hence, main focus of study is on the dynamic displacement and fundamental mode of vibration of the bridge.

To study the response of the bridge with respect to varying the velocity of the vehicle, the peak mid-span vertical displacement and acceleration response of the bridge are plotted in Fig. 3, against varying the velocity of the vehicle. It is observed that the displacement and acceleration responses of the bridge become excessive at the first critical train velocity, which confirms the agreement of Eq. (30), that for higher values of l, the critical velocities are lower and the corresponding response peaks are not large and do not require to be controlled practically. The time history of uncontrolled displacement and acceleration responses at mid-span of the bridge along the vertical translational DOF for the first critical velocity of Japanese SKS train moving over the bridge are shown in Fig. 4. It shows that the dynamic responses of the bridge become excessive when the vehicle moving over it runs at critical velocities, leading to the resonating conditions. The critical velocity depends on the fundamental frequency of the bridge and the axle spacing of vehicles.

Table 1 Properties of the THSR bridge

Properties	Values
Length of span, L (m)	30.0
Elasticity modulus, E_m (N/m²)	2.83×10^{10}
Moment of inertia, I_m (m⁴)	7.84
Mass per unit length, m (kg/m)	41.74×10^3
Modal damping ratio, ξ (%)	2.5
Fundamental natural frequency, ω_1 (rad/s)	25.3

Table 2 Properties of model of Japanese SKS train

Properties	Values
Length of train, L_v (m)	402.1
Number of axles, n_w (number)	16
Axle distance, x_v (m)	25.0
Axle load, P (N)	552.0×10^3
Critical velocity, $(v_c)_{l=1}$ (m/s)	101.0

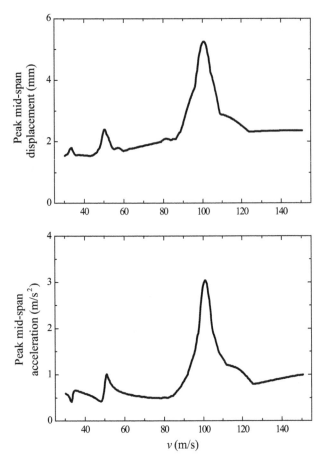

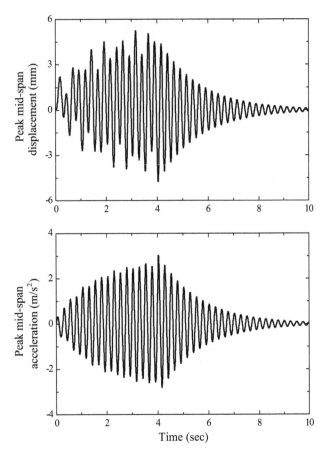

Fig. 3 Peak mid-span responses of bridge against vehicle velocities

Fig. 4 Time history responses at the mid-span of bridge for the first resonant vehicle velocities

Controlled response of bridge and optimization of parameters

To control response of the bridge, different TMFD systems are concentrated at the mid-span of the bridge or distributed at an equal interval along the length of the bridge. These systems perform effectively only when the appropriate value of controlling parameters, namely, frequency spacing, tuning frequency ratio, damper slip force R_f and number of TMFD units in an MTMFD, is selected. In the case of inaccurate selection of these parameters, the system may under-perform and the responses may not reduce effectively up to a desirable value. Hence, the optimization of the parameters of TMFD systems is a very important criterion for its effective functioning. In this study, the parameters are optimized to reduce the peak mid-span displacement response of the bridge to its minimum value.

The mass ratio of all the TMFD units of MTMFD systems is kept the same as that of TMFD system. However, the criteria of optimization of parameters of all the TMFD systems (TMFD, concentrated MTMFD and distributed MTMFD) are the same, i.e., the minimization of peak mid-span displacement response of the bridge. After the

selection of the number of TMFD units in MTMFD systems and the mass ratio of TMFD system, the maximum responses of the bridge subjected to multi-axle vehicles are studied.

Optimization of parameters for concentrated TMFD

The variation of the optimum parameters, β^{opt}, f^{opt}, and R_f^{opt} against the number of TMFD units in an MTMFD system concentrated at the mid-span of the bridge is shown in Fig. 5. It is observed that the optimum frequency spacing, β^{opt}, increases sharply with the increase in the number of the TMFD units and beyond certain numbers of TMFD units, it increases gradually. Similarly, the optimum frequency ratio, f^{opt}, increases with the increase in the number of TMFD units and remains constant after a certain number of TMFD units. The optimum value of normalized slip force, R_f^{opt}, of MTMFD system is much lower than single TMFD system. It is visible from the plot that the optimum values of R_f^{opt} reduce sharply with the increase in the number of TMFD units, up to a certain number of TMFD units, and beyond this number, the response curve becomes flatter. It is also observed from the response plot that with

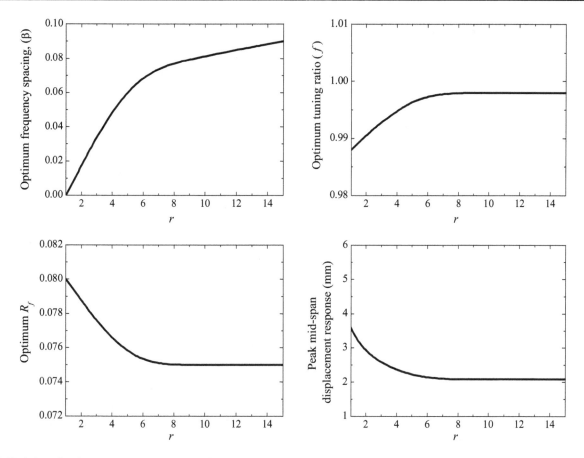

Fig. 5 Variation of optimum parameters and peak mid-span response of bridge against the number of TMFD units concentrated at the mid-span of bridge

the increasing number of TMFD units, the peak response reduces monotonically up to a certain number of TMFD units and after that the reduction becomes insignificant. In the present case, the increase in the number of TMFD units beyond 5 will make the reduction of dynamic displacement response of THSR bridge practically insignificant. Hence, in this present context, the MTMFD system is selected with five TMFD units. Thus, an optimum value of controlling parameters, such as, β^{opt}, f^{opt}, and R_f^{opt} varies with respect to the number of TMFD units at which system can perform effectively.

The peak mid-span displacement response of the bridge is plotted against various controlling parameters of TMFD systems in Fig. 6 for the mass ratio of 2 % for TMFD and concentrated MTMFD system (containing five numbers of TMFD units). In Fig. 6a, the response of system is plotted against the varying values of frequency spacing, keeping the optimum value of tuning frequency ration and R_f constant. Similarly, in Fig. 6b, the optimum value of frequency spacing and R_f is kept constant for each TMFD system and the response of the system is plotted against tuning frequency ratio. In addition, Fig. 6c shows the response of system against varying the values of R_f for both

the TMFD systems, keeping the optimum value of tuning frequency ratio and frequency spacing constant. It is observed from Fig. 6 that the MTMFD system is sensitive to the frequency spacing. In addition, peak mid-span displacement response of the bridge reduces to its minimum value at a particular value of tuning frequency ratio, frequency spacing and R_f for each TMFD unit. The optimum parameters of concentrated MTMFD system are summarized in Table 3. Thus, MTMFD system is sensitive to the frequency spacing. At the optimum value of controlling parameters, namely frequency spacing, tuning frequency ratio and R_f and optimum number of TMFD units, the response of the bridge reduces to its minimum value.

Optimization of parameters for distributed MTMFD

To optimize the parameters of distributed MTMFD, the TMFD units of MTMFD are distributed at an equal interval of 2 and 5 m, respectively, along the length of the bridge. For a fixed number of TMFD units distributed at fixed interval, the parameters can be optimized in a similar way as optimized for concentrated MTMFD system with the

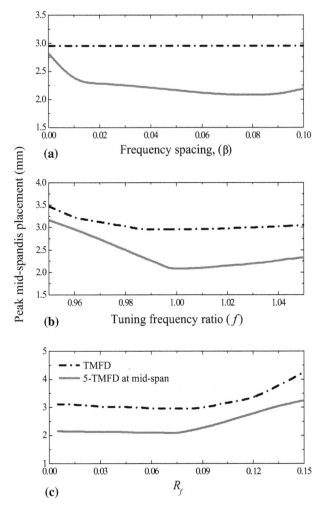

Fig. 6 Peak mid-span displacement responses of bridge against different TMFD parameters

same governing criteria of minimization of peak mid-span displacement response of the bridge. In this study, the distributed MTMFD system is always placed symmetrically with respect to the mid-span of the bridge with the heaviest TMFD unit always at the mid-span and the mass of TMFD unit decreasing with its distance from the mid-span on either side, hence, only odd number (minimum three) of TMFD units are considered to comprise the

distributed MTMFD system. The maximum number of TMFD units which can be used depends on the length of the bridge and the interval of TMFD units. Figure 7 shows that the nature of variation of optimum parameters with respect to the number of TMFD units composing the distributed MTMFD system is very similar to that of concentrated MTMFD system shown in Fig. 5, if the TMFD units are distributed at a fixed interval of 2 m. The optimum frequency spacing and optimum tuning frequency ratio increase with the increasing number of TMFD units. The optimum normalized slip force, R_f^{opt}, of MTMFD system is lower than TMFD system, as it reduces with the increase in the number of TMFD units up to a certain number of TMFD units and after that the response curve becomes flatter. Similar to the case of concentrated MTMFD system, the peak mid-span displacement response of the bridge reduces monotonically with the increase in the number of TMFD units of distributed TMFD system up to a certain number of TMFD units and beyond this number, and the rate of response reduction becomes practically insignificant.

Like concentrated MTMFD, the optimum parameters of distributed MTMFD system are summarized in Table 3. The responses shown in Fig. 7d are the minimized peak mid-span displacement responses of the bridge, considering optimum values of controlling parameters corresponding to each number of TMFD units. Thus, the optimum frequency spacing of MTMFD system increases with the number of TMFD units of MTMFD system. The optimum tuning frequency ratio increases with the number of TMFD units. The optimum R_f reduces with the increase in the number of TMFD units. The optimum R_f of TMFD is much higher than that of MTMFD system. The response of the bridge decreases with the increase in the number of TMFD units of a distributed MTMFD system up to a certain number of TMFD units and after that the reduction of response becomes practically insignificant which shows an optimum number of TMFD units in distributed MTMFD exists. For the present study, the optimum number of TMFD units is selected as 5 to compose distributed TMFD system.

Table 3 Optimum parameters of TMFD systems

TMFD system	Mass ratio (μ)	Frequency spacing (β)	Tuning frequency ratio (f)	Normalized friction force (R_f)
TMFD	0.02	–	0.988	0.08
5-TMFD units concentrated at mid-span of bridge	0.02	0.07	0.998	0.075
5-TMFD units distributed at 2 m interval under bridge	0.02	0.03	1.0	0.07

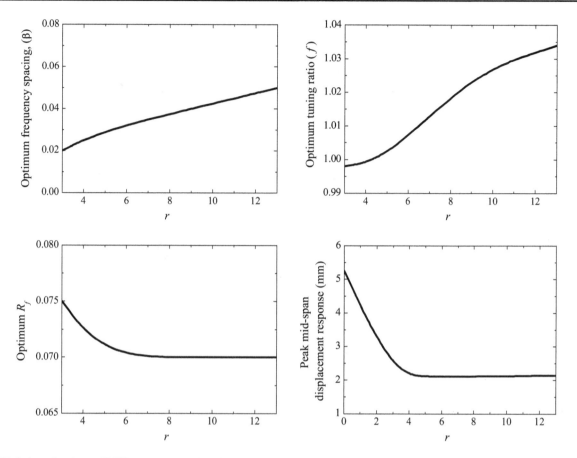

Fig. 7 Variation of optimum TMFD parameters and peak mid-span response of bridge against the number of TMFD units distributed at a fixed interval of 2 m along the length of the bridge

Effect of distribution of TMFD units along length of bridge

The effect of distribution of TMFD units along the length of the bridge is studied in Fig. 8. For this purpose, five TMFD units are distributed with equal interval of different values along the length of the bridge. It is observed that with the increase in the interval of the TMFD units, the peak mid-span displacement and acceleration response of the bridge increase. Thus, the MTMFD system consisting of five TMFD units is most effective when all the TMFD units are concentrated at the mid-span of the bridge, i.e., when the interval is zero, and its efficiency in reduction of bridge responses decreases with the increase in interval of the TMFD units. Thus, the MTMFD is most effective, if all the TMFD units are concentrated at the mid-span. Wang et al. (2013) has shown that the maximum response of the bridge can be noticeably suppressed if VED is installed at the midpoint of each span of the bridge, and the same is also confirmed for MTMFD system having all the TMFD units concentrated at the mid-span of the bridge.

Effect of mass ratio

The effect of mass ratio on the performance of different TMFD systems is studied in Fig. 9 by plotting the peak mid-span displacement and acceleration response of the bridge against the varying mass ratio for different TMFD systems, namely, TMFD, concentrated MTMFD and MTMFD distributed at 2 m interval. It is observed that the peak mid-span responses of the bridge decreases with an increase in the mass ratio of all the TMFD systems up to a certain value of mass ratio and after that it gradually increases. In addition, the reduction is maximum for concentrated MTMFD system and minimum for TMFD system. Thus, similar to the optimum controlling parameters, an optimum value of mass ratio exists for all the TMFD systems, at which the response reduction of the bridge is maximum. In addition, the distributed MTMFD system having optimized controlling parameters with appropriately distributed TMFD units within a certain interval can be more effective than a TMFD.

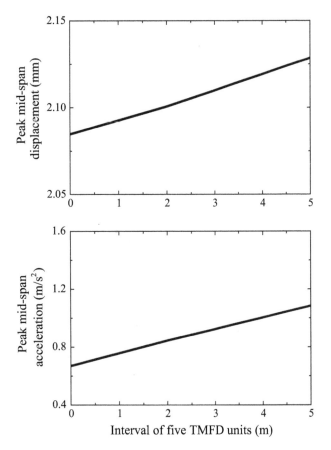

Fig. 8 Peak mid-span displacement and acceleration response of bridge for varying interval of TMFD units of MTMFD system having five-TMFD units distributed along the length of the bridge at an equal interval

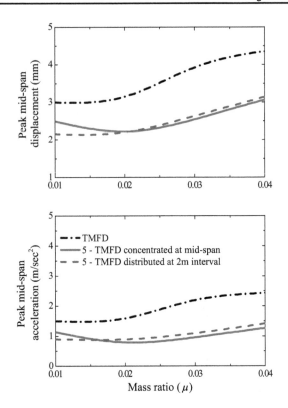

Fig. 9 Peak mid-span displacement and acceleration response of bridge against mass ratios of different TMFD systems

Table 4 Mass distribution of TMFD units of different TMFD systems

Mass ratio	Distribution of mass of TMFD units (kg)	
	TMFD	Concentrated and distributed MTMFD systems
0.02	25,044	5369, 5179, 5000, 4829, 4667

Resonant response control of the bridge with different TMFD systems

In this section, the effectiveness of TMFD, concentrated MTMFD and distributed MTMFD systems in the reduction of excessive dynamic responses of the bridge under the excitations induced by moving multi-axle vehicle is highlighted. The functioning of different TMFD systems is compared with the help of figures. The distribution of mass of TMFD and TMFD units of MTMFD system are mentioned in Table 4. It is observed from Figs. 10 and 11 that optimized TMFD systems are very effective in the reduction of resonant displacement as well as acceleration responses of bridge under excitations induced by moving multi-axle vehicles. All the TMFD systems are effective at, or very near to the vicinity of resonant zone only, and apart from this zone, they are not so effective. This is because the TMFD systems are sensitive to the frequency change, and in this study, the TMFD system are tuned to the resonating frequency of the bridge. Furthermore, the comparative study of performance of all the TMFD systems shows that

at resonating condition, the maximum reduction of mid-span displacement and mid-span acceleration responses is achieved with the use of optimized concentrated MTMFD systems, i.e., placing all the TMFD units at the mid-span of the bridge. It is also observed that when the same number of TMFD units is distributed at a fixed interval of 2 and 5 m, respectively, along the length of the bridge, it becomes less effective than a concentrated MTMFD system, but if the frequencies of the TMFD units are optimized efficiently, and they are placed at an interval within a certain limit then this distributed system becomes more effective than a TMFD system in the reduction of resonant responses of the bridge. In this present case, it is observed that an optimized distributed MTMFD system consisting of five TMFD units is more effective than an optimized TMFD system when the interval of the TMFD units is 2 m, but at the same time, it is less effective when the interval becomes 5 m. Thus, all the optimized TMFD systems are very effective in reducing the resonant displacement as

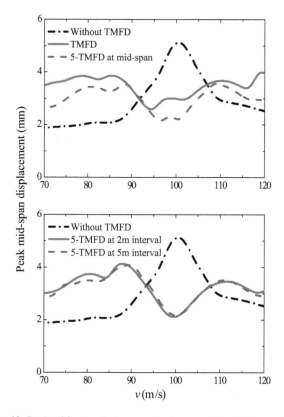

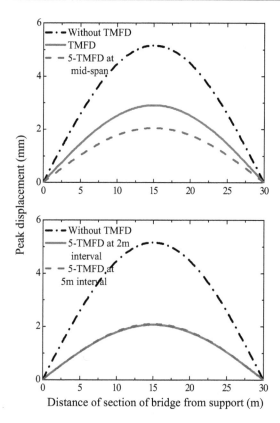

Fig. 10 Peak mid-span displacement response of the bridge against the varying vehicle velocities with different TMFD systems

Fig. 12 Peak displacement responses at different sections of bridge along its length with different TMFD systems

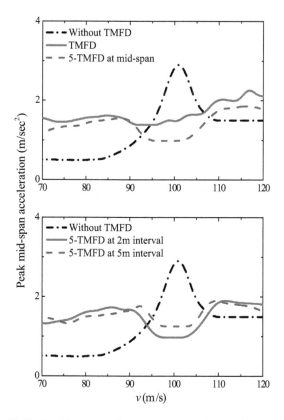

well as acceleration responses of the bridge at or very near to the vicinity of resonant zone and a very part from this zone, they are not so effective. An optimized concentrated MTMFD system is the most effective system in reducing the resonant displacement and acceleration response of the bridge. When the number of TMFD units of an MTMFD system is distributed at a fixed interval along the length of the bridge, it becomes less effective than a concentrated MTMFD system, but if the frequencies of TMFD units are optimized efficiently, it becomes more effective than a TMFD system in reducing the resonant responses of bridge.

Furthermore, the performance of all the optimized TMFD systems is studied at different sections along the length of the bridge and is represented in Figs. 12 and 13. It is observed that all the TMFD systems are effective in reducing displacement as well as acceleration responses of the bridge at all the considered sections along its length, with the reduction being maximum at the mid-span of the bridge. It is also observed that the optimized MTMFD system consisting of five TMFD units concentrated at the mid-span of the bridge significantly reduces the response of the bridge at all the considered sections, under the influence of moving multi-axle vehicle. The reduction with the optimized TMFD system is less than the reduction with

Fig. 11 Peak mid-span acceleration response of the bridge against the varying vehicle velocities with different TMFD systems

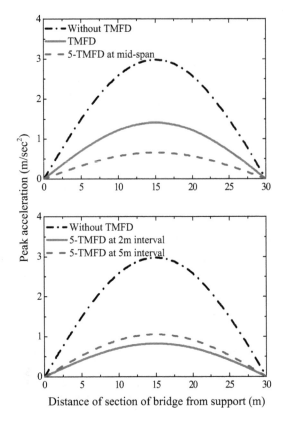

Fig. 13 Peak acceleration responses at different sections of bridge along its length with different TMFD systems

optimized MTMFD (consisting of five TMFD units) system distributed at 2 and 5 m interval. Thus, all the optimized TMFD systems are effective in reducing the resonant displacement and acceleration responses at all the sections of the bridge, with the maximum reduction at the mid-span of the bridge.

The mid-span displacement and acceleration responses of the bridge with installation of different TMFD systems are listed in Table 5. It shows that all the TMFD systems appreciably reduces the mid-span displacement and acceleration response of the bridge. MTMFD system which is concentrated at the mid-span of the bridge can reduce displacement response up to 60.28 % and acceleration response up to 77.87 %, respectively.

Conclusions

The performance of MTMFD in controlling the undesirable response of the bridge is investigated. Simplified model of THSR bridge and Japanese SKS train is prepared, and the velocities of train causing undesirable resonant responses in the bridges are considered. To suppress excessive resonant responses of the bridge, different TMFD systems are employed under the bridge. The optimum parameters of different TMFD systems are found out with the criteria of minimization of peak mid-span displacement response of the bridge. The effect of distribution of TMFD units along the length of the bridge for MTMFD system is also investigated. On the basis of trends of results obtained, the following conclusions are drawn:

1. The dynamic responses of the bridge become excessive when the vehicle moving over it runs at critical velocities, leading to the resonant conditions. The critical velocity depends on the fundamental frequency of the bridge and the axle spacing of the vehicles.
2. An optimum value of controlling parameters, such as, β^{opt}, f^{opt}, and R_f^{opt} varies with respect to the number of TMFD units at which system can perform effectively.
3. MTMFD system is sensitive to the frequency spacing. At the optimum value of controlling parameters, namely, frequency spacing, tuning frequency ratio and R_f and optimum number of TMFD units, the response of the bridge reduces to its minimum value.
4. The MTMFD is most effective if all the TMFD units are concentrated at the mid-span.
5. An optimum value of mass ratio exists for all the TMFD systems, at which the response reduction of the bridge is maximum.
6. All the optimized TMFD systems are very effective in reducing the resonant displacement as well as acceleration responses of the bridge at or very near to the vicinity of resonant zone and very apart from this zone, they are not so effective. An optimized concentrated MTMFD system is the most effective system in reducing the resonant displacement and acceleration response of the bridge.

Table 5 Peak mid-span displacement and acceleration responses of bridge with different TMFD systems

TMFD condition	Displacement (mm)	Acceleration (m/s²)	Displacement reduction (%)	Acceleration reduction (%)
Without TMFD	5.249	3.028	–	–
With TMFD at mid-span	2.955	1.431	43.70	52.74
With 5-TMFD units at mid-span	2.085	0.670	60.28	77.87
With 5-TMFD units at 2 m interval	2.101	0.846	59.97	72.06
With 5-TMFD units at 5 m interval	2.128	1.084	59.46	64.20

7. When the number of TMFD units of a MTMFD system is distributed at a fixed interval along the length of the bridge, it becomes less effective than a concentrated MTMFD system, but if the frequencies of TMFD units are optimized efficiently, it becomes more effective than a TMFD system in reducing the resonant responses of the bridge.

8. All the optimized TMFD systems are effective in reducing the resonant displacement and acceleration responses at all the sections of the bridge, with the maximum reduction at the mid-span of the bridge.

9. This study can be continued in the future to investigate the effect of track irregularity, lateral displacement, 3D effect of bridge-vehicle system on the response of the bridge. In addition, the performance of continuous span of the bridge with some rigid supports can be studied, using different TMFD systems.

References

Antolin P, Zhang N, Goicolea JM, Xia H, Astiz MA, Oliva J (2013) Consideration of nonlinear wheel-rail contact forces for dynamic vehicle-bridge interaction in high-speed railways. J Sound Vib 332:1231–1251

Bhaskararao AV, Jangid RS (2006) Seismic analysis of structures connected with friction dampers. Eng Struct 28:690–703

Chen YH, Lin CY (2000) Dynamic response of elevated high speed railway. J Bridge Eng ASCE 5:124–130

Cheng YS, Au FTK, Cheung YK (2001) Vibration of railway bridges under a moving train by using bridge-track-vehicle element. Eng Struct 23:1597–1606

Chopra AK (2003) Dynamics of structures theory and applications to earthquake engineering. Prentice Hall, New Delhi

Constantinou M, Mokha A, Reinhorn A (1990) Teflon bearing in base isolation. Part II: modeling. J Struct Eng ASCE 116:455–474

Jianzhong L, Mubiao S, Lichu F (2005) Vibration control of railway bridges under high-speed trains using multiple TMDs. J Bridge Eng ASCE 10:312–320

Ju SH, Lin HT (2003) Resonance characteristics of high-speed trains passing simply supported bridges. J Sound Vib 267:1127–1141

Kwon HC, Kim MC, Lee IW (1998) Vibration control of bridges under moving loads. Comput Struct 66:473–480

Li Q, Xu YL, Wu DJ, Chen ZW (2010) Computer-aided nonlinear vehicle-bridge interaction analysis. J Vib Control 16(12):1791–1816

Lin CC, Wang JF, Chen BL (2005) Train induced vibration control of high speed Railway bridges equipped with multiple tuned mass dampers. J Bridge Eng ASCE 10(4):398–414

Moghaddas M, Esmailzadeh E, Sedaghati R, Khosravi P (2012) Vibration control of Timoshenko beam traversed by moving vehicle using optimized tuned mass damper. J Vib Control 18(6):757–773

Nasiff HH, Liu M (2004) Analytical modeling of bridge-road-vehicle dynamic interaction system. J Vib Control 10(2):215–241

Pisal AY, Jangid RS (2015) Seismic response of multi-story structure with multiple tuned mass friction dampers. Int J Adv Struct Eng 7:81–92

Shi X, Cai CS (2008) Suppression of vehicle-induced bridge vibration using tuned mass damper. J Vib Control 14(7):1037–1054

Wang JF, Lin CC, Chen BL (2003) Vibration suppression for high-speed railway bridges using tuned mass dampers. Int J Solids Struct 40:465–491

Wang YJ, Yau JD, Wei QC (2013) Vibration suppression of train-induced multiple resonant responses of two-span continuous bridges using VE dampers. J Mar Sci Technol 21(2):149–158

Wen YK (1976) Method for random vibration of hysteretic systems. J Eng Mech Div ASCE 102(2):249–263

Yang YB, Yau JD, Hsu LC (1997) Vibration of simple beams due to trains moving at high speeds. Eng Struct 19:936–944

10

Investigation of the nonlinear seismic behavior of knee braced frames using the incremental dynamic analysis method

Mohammad Reza Sheidaii[1] · Mehrzad TahamouliRoudsari[2] · Mehrdad Gordini[2]

Abstract In knee braced frames, the braces are attached to the knee element rather than the intersection of beams and columns. This bracing system is widely used and preferred over the other commonly used systems for reasons such as having lateral stiffness while having adequate ductility, damage concentration on the second degree convenience of repairing and replacing of these elements after Earthquake. The lateral stiffness of this system is supplied by the bracing member and the ductility of the frame attached to the knee length is supplied through the bending or shear yield of the knee member. In this paper, the nonlinear seismic behavior of knee braced frame systems has been investigated using incremental dynamic analysis (IDA) and the effects of the number of stories in a building, length and the moment of inertia of the knee member on the seismic behavior, elastic stiffness, ductility and the probability of failure of these systems has been determined. In the incremental dynamic analysis, after plotting the IDA diagrams of the accelerograms, the collapse diagrams in the limit states are determined. These diagrams yield that for a constant knee length with reduced moment of inertia, the probability of collapse in limit states heightens and also for a constant knee moment of inertia with increasing length, the probability of collapse in limit states increases.

Keywords Knee brace · Incremental dynamic analysis · Behavior factor · Collapse diagram

Introduction

Structures designed to resist moderate and frequently occurring earthquakes must have sufficient stiffness and strength to control deflection and to prevent any possible damage. However, it is inappropriate to design a structure to remain in the elastic region, under severe earthquakes, because of the economic constraints. The inherent damping of yielding structural elements can advantageously be utilized to lower the strength requirement, leading to a more economical design. This yielding usually provides the ductility or toughness of the structure against the sudden brittle type structural failure. It is desirable to devise a structural system that combines stiffness and ductility, in the most effective manner without excessive increase in the cost. In building seismic design codes, two main goals are aimed for. First, during weak and average Earthquakes the structure should have sufficient strength and stiffness to prevent structural damages and to control the displacement. Second, during strong Earthquakes, the structures must have the ability to absorb energy properly and display suitable ductility. Stiffness and ductility are two elements that are frequently at odds, thus it is desirable for the structural system to have rational and sensible balance between the two (Balendra and Sam 1990, 1991a).

At present, in designing seismic resistant steel buildings, systems such as moment frames, concentrically braced frames and eccentrically braced frames are widely used. Moment resisting frames show a good ductility due to the bending yield of the beam but its stiffness in low. Concentrically braced frames have a high stiffness but due to

✉ Mehrzad TahamouliRoudsari
Tahamouli@iauksh.ac.ir

[1] Department of Civil Engineering, College of Engineering, Urmia University, Urmia, Iran

[2] Department of Civil Engineering, Kermanshah Branch, Islamic Azad University, Kermanshah, Iran

the buckling of the compressional brace, it displays a low ductility. To overcome the problem of stiffness in moment resisting frames and also the problem of low ductility in concentrically braced frames, Hjelmstad and Popov (1983) proposed the eccentric brace. By considering the appropriate amount of eccentricity, the system contains sufficient stiffness and through the shear and bending yield of the intermediary beam, the ductility will be provided. This system has fitting ductility and stiffness but to supply its ductility, the intermediary beam, which is one of the main elements of the structure, must yield and that results in severe damage to the ceiling which makes the post-earthquake repairing difficult (Balendra and Sam 1991b, 1994).

Ochao (1986) proposed a new system which was later modified by Balendera (1990, 1991a, 1994). In this system which is called the (KBF), the tag end of the diagonal braces are attached to a skewed (knee) element which itself is attached to the beam and the column or the beam and the support, rather than being annexed to the intersection of the beam and the column. In this system the knee element remains in an elastic state during small earthquake and they yield before the main members of the structure in strong earthquakes, and this causes the energy to be dissipated without the lateral strength being attenuated. In this system, the damages brought about by the earthquake will be concentrated on the knee members which are not key structural members and can be replaced and repaired after the earthquake (William and Denis 2004).

In recent years, different studies have been conducted on determining dimensions, shape, characteristics and other optimal parameters of these systems in order to achieve the most desirable stiffness—ductility combinations. The later experimental work on directly knee braced model frames by Zahrai and Jalali (2014) and Sutat et al. (2011) also showed the applicability of this method. Knee braces were

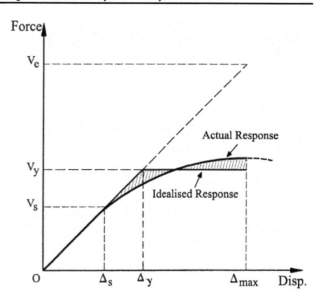

Fig. 1 Parameters used in calculation of ductility, overstrength and performance factor

used in the past for wind-resistant design and have been recently explored in various forms for seismic applications (Inouel et al. 2006; Lee and Bruneau 2005). The design of the proposed KBMF structural system is based on a capacity-design concept that results in ductile behavior. For this system, the frames are designed so that the knee braces will yield and buckle under seismic loads; this is followed by plastic hinging of beams at the ends of the beam segments outside the knee portions. All inelastic activities are directed away from the critical areas, decreasing the dependence of the performance on the material and quality of the welded joints. In this study, also, in the interest of completing the former studies, the principles of incremental dynamic analysis have been employed.

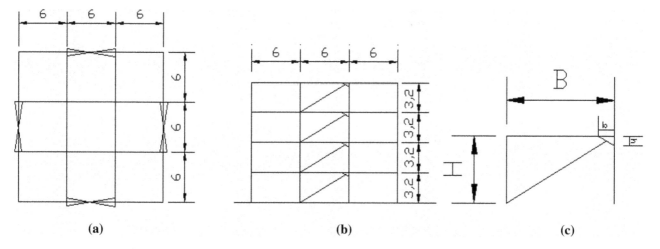

(a)   **(b)**   **(c)**

Fig. 2 Geometrical specifications of the building. **a** Plan of building. **b** View of building. **c** Knee braced

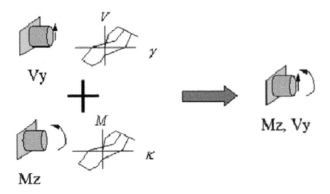

Fig. 3 Section used for knee element in OpenSees software (McKenna et al. 2007)

Estimating the behavior factor

Most of the design codes decrease the base shear by dividing the elastic earthquake factor over the larger number of a unit called "the behavior factor". The below equation can be used as a low approximation of the behavior factor (R) (Chopra 1998).

$$R = R_\mu \times R_S \times Y \tag{1}$$

where R_μ is the ductility factor of the system and presents the energy dissipation capacity. R_S is the excess strength factor of the system and Y is the allowable stress factor. These parameters will be determined through Eqs. 2, 3 and 4 (Fig. 1) using the capacity spectra diagram, where V_d is the designing shear, V_s is the shear that corresponds to the development of the first plastic joint, V_{ym} is the base shear established upon the maximum inelastic displacement and V_e is the maximum seismic need for the elastic response, F_1, F_2 and F_3 are respectively the ratio of

actual stress over the nominal stress, the effect of the speed of loading in increasing the yield stress and the effect of nonstructural (Chopra 1998; Jinkoo 2005).

$$R_S = F_1 \times F_2 \times F_3 \times R_{SO} \tag{2}$$

$$R_\mu = V_e / V_{ym} \tag{3}$$

$$R_{so} = V_{ym} / V_s \tag{4}$$

In this study the behavior factor is obtain through the multiplication of R_μ and R_S, and R_{so} is equal to V_{ym}/V_d.

The subjected models

To assess the stiffness, the factor of ductility, the factor of excess strength, the behavior factor and the probability of failure of knee braced systems, a building like the one depicted in Fig. 2 with the below characteristics has been studied.

The steel building has three bays along their two main directions and the number of the stories is 4 and 8. The length of the bays and the height of the stories are 6 and 3.2 meters, respectively. The building is braced in both of its lateral sides to resist earthquake. The weight of each story is 274 ton and it's been assumed that the building are constructed in Tehran with very high earthquake risk, soil type 3 (BHRC: Standard No. 2800, 2005) whose shear velocity is approximately between 175 and 375 m/s and residential application. The connection between the beam and the column has been considered as a joint. These buildings have designed in accordance with the AISC-ASD design code (2005). The specifications of the knee brace is presented in Fig. 2a and is described as follow. The knee element is placed on one side and at the top of the brace. The type of connection between the knee and the beam and

Table 1 Specifications of the accelerograms use in the incremental dynamic analyzes

No.	Earthquake	Station	Moment Magnitude	Distance (km)
1	Friuli-Italy 1976	Codroipo	6.5	33.4
2	Cape Mendocino	Eureka	7.01	41.97
3	Cape Mendocino	Eureka	7.01	41.97
4	Imperial Valley	El Centro	6.5	18.20
5	Imperial Valley	Calipatria	6.5	23.17
6	Loma Prieta	Gilory	6.9	16.1
7	Loma prieta	Palo Alto	6.9	36.3
8	Northridge	La Saturn	6.7	30
9	Chi Chi	CWB Chy002	7.62	24.21
10	Chi Chi	CWB Chy004	7.62	24.21
11	Duzce	Yarimca	7.14	97.53
12	Kocaeli	Ambarli	7.51	69.62
13	Kocaeli	Ambarli	7.51	69.62
14	Kobe	Kakogava	6.9	22.5
15	Kobe	Kakogava	6.9	22.5

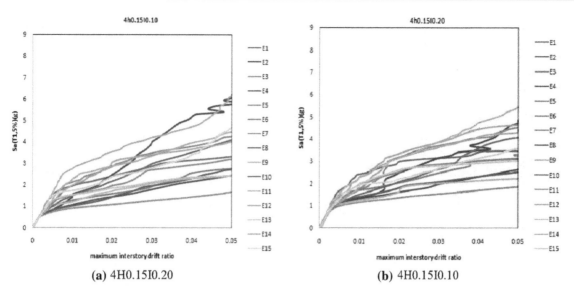

(a) 4H0.15I0.20   **(b)** 4H0.15I0.10

Fig. 4 The results of the incremental dynamic analysis of 4 story frames

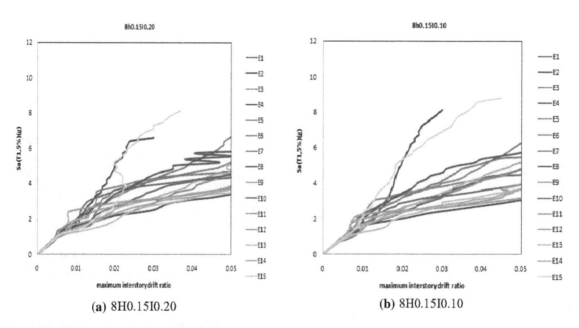

(a) 8H0.15I0.20   **(b)** 8H0.15I0.10

Fig. 5 The results of the incremental dynamic analysis of 8 story frames

column and the connection between the brace and the knee are considered rigid and joint, respectively. The knee element is parallel to the other hypotenuse of the frames in such a way that the ratios b/h and B/H are equal and the along of the bracing member goes through the point at which the beam and the column are connected (Mofid and Khosravi 2000; Mofid and Lotfollahi 2006). Parametrical studies conducted on knee brace frames for different lengths and moment of inertia of the knee member are shown below..

$h/H = 0.0, 0.05, 0.10, 0.15, 0.20, 0.25, 0.30$

$I_K/I_C = 0.10, 0.15, 0.20, 0.25, 0.30$

where h, H, I_C and I_k are the vertical length of the knee element, the height of the story, the moment of inertia of the column and the moment of inertia of the knee respectively.

The frames are generally named using the format "xBHaIb" in which x, a and b are the number of stories, the ratio of h/H and the ratio of I_C/I_K, respectively. For example 8BH0.15I0.25 would means the frame with 8 story knee brace with the ratios of $h/H = 0.15$ and $I_k/I_c = 0.25$, the diagonal brace, also, is shown using "gh".

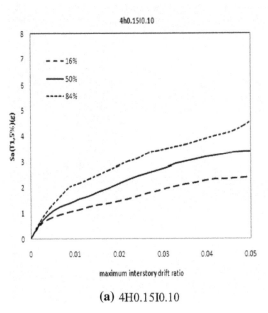

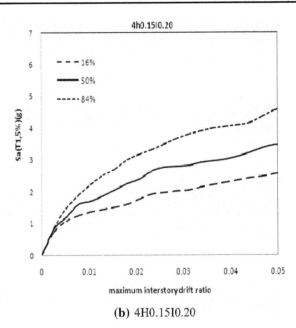

(a) 4H0.15I0.10   **(b)** 4H0.15I0.20

Fig. 6 A summary of the IDA results of 4 story buildings

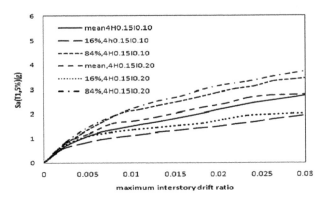

Fig. 7 Comparing the summary of the IDA results of 4 story buildings

Incremental dynamic analysis

The incremental dynamic analysis has been carried out using the software opensees. For beams, columns and the bracing members, the fiber section for the nonlinear beam column element has been used and for the knee members, shear yield specifications to be exertable the section aggregator nonlinear beam column element has been used (Fig. 3). For the material specifications, the steel 02 material with the yield stress of 2400 kg/cm^2 has been used and the modulus of elasticity and the post yield stiffness are considered equal 2.1e6 kg/cm^2 and 2 %, respectively. The allowable drift according to the Iranian standard design code (2005) and the UBC for the frames of the both building are considered equal to 0.025. one of the following has been considered as the collapse mode: the drift exceeding the allowable limit or the buckling of the brace. The loading was continued until one of these two collapse modes were observed in the shear–displacement diagram.

The incremental dynamic analysis (IDA) has been employed to analyze the frames. The intensity measure (IM) has been taken into account in the spectral acceleration analysis of the first vibration mode of the structure [s_a (T_1)], and the maximum inter story drift ratio (MIDR) has been considered as the damage measure (DM). For these analyses, 15 accelerograms whose specifications are demonstrated in Table 1 have been used.

Result of dynamic analysis of 4 and 8 story frames with constant knee length and variable moment of inertia

The result of dynamic analysis of 4 and 8 story frames with constant knee length and variable moment of inertia are presented in Figs. 4 and 5. From linear responses to strictly nonlinear responses of the structure are observable in these figures. The structures response to some of the accelerograms is intensive and not so severe to some others, i.e., for a very small increase in the spectral acceleration of the accelerogram, the maximum intensity measure of the drift expands rapidly (softening), meanwhile, for some other accelerogram by drastically increasing the intensity measure, the stiffness measure wont experience any significant change (hardening), and all these responses in the incremental dynamic analysis are not unexpected. In fact in these diagrams, we can observe all of the probable responses during the future earthquakes.

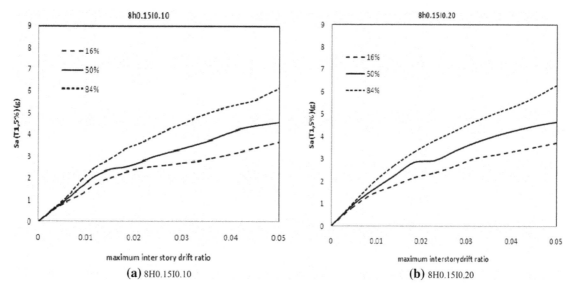

(a) 8H0.15I0.10   **(b)** 8H0.15I0.20

Fig. 8 A summary of the IDA results of 8 story buildings

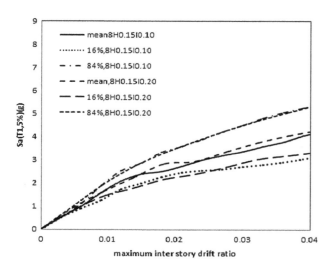

Fig. 9 Comparing the summary of the IDA results of 8 story buildings

In Figs. 6 and 7, summary of the IDA's results is presented which includes the average curve and the 16 and 84 % quintiles. The quintile, actually, show the dispersion of the results of the analysis.

The summary of the IDA curve makes the interpretation of the countless results of the analysis possible. The 50 % curve shows the average dynamic response of the structure to earthquakes and yields the maximum drift of the stories in terms of the magnitude of the earthquake. The 16 and 84 % diagrams show the changeability of this average value. It is worth to point out that although the incremental dynamic analysis is the most powerful and reliable method at hand to analyze a structure subjected to seismic loads, still it is accompanied with excessive uncertainties in the

exact response of the structure. To put that in perspective's pace as it is seen, the distance between the quintiles of the results of the analysis and the average diagram is considerable and this phenomenon enhances in extremely nonlinear domains, which makes the practical use of the results difficult. Despite all of that, the aforementioned results have the lowest deviation from the intensity and damage measures. For this reason, it seems that because of the above short comings, the path will be opened to exacting the incremental dynamic analysis, e.g., assessing the sensitivity of the results to different intensity parameters.

As it can be seen in the diagrams that compare the Figs. 8 and 9, the average diagrams and the quintiles overlap on a certain drift which in fact shows the elastic limit and the linear behavior of the structure. Prior to this drift, the structures response to different accelarograms will not go through any kind of change and the displacement is proportionate to the force and in this range, even the hooks low can be implemented to obtain the stiffness of the structure. ($F = KX$). After this drift the quintile diagrams deviate from the average diagram, but still the average diagrams illustrate equal responses.

This part shows the development of linear behavior in the structures, but still, inducing the moment of inertia has not affected the behavior of the structure. After this point, the average diagram of the frames with low knee moment of inertia will be under the frame with higher moment of inertia and this shows that reduction in the moment of inertia causes the frames strength to decrease and that for a slight increase in the earthquake force, the frame experiences a larger drift.

In order to more accurately show the behavior of the structure along the changes of the drifts have been used and

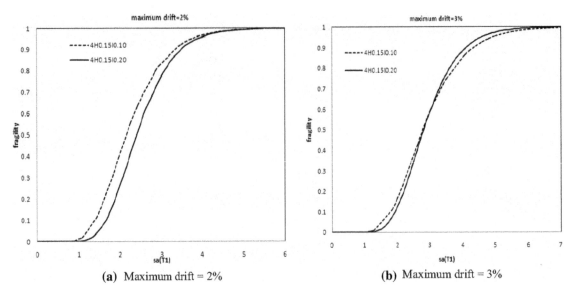

(a) Maximum drift = 2% **(b)** Maximum drift = 3%

Fig. 10 The collapse diagrams to surpass the 2 and 3 % drifts of 4 story frames

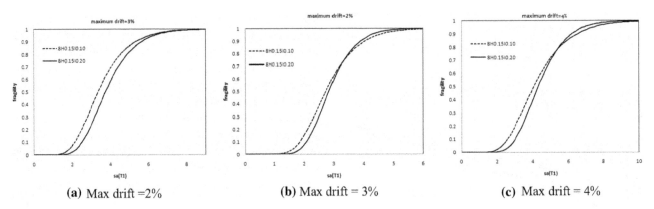

(a) Max drift =2% **(b)** Max drift = 3% **(c)** Max drift = 4%

Fig. 11 The collapse diagrams to surpass the 2, 3 and 4 % drifts of 8 story frames

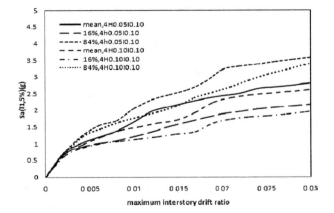

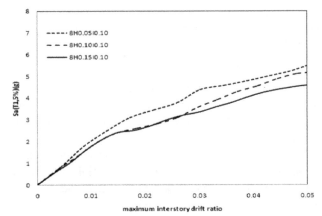

Fig. 12 Comparing the summary of the IDA results of 4 story buildings

Fig. 13 Comparing the summary of the IDA results of 8 story buildings

the collapse diagrams of 2, 3 and 4 % the frames in the aforementioned drifts are compared. The collapse diagrams can be observed in Figs. 10 and 11.

As it can be seen from the collapse diagrams, for the mentioned drifts with low knee moment of inertia are above those with high knee moment of inertia and so the

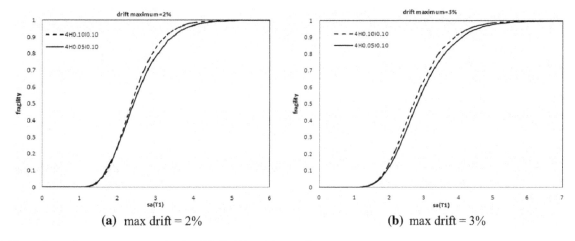

(a) max drift = 2%   **(b)** max drift = 3%

Fig. 14 The collapse diagrams to surpass the 2 and 3 % drifts of 4 story frames

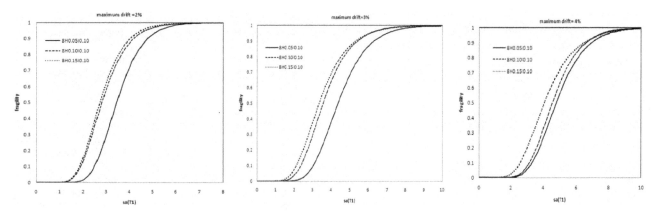

Fig. 15 The collapse diagrams to surpass the 2, 3 and 4 % drifts of 8 story frames

probability of failure of frames with low knee moment of inertia is higher than those with low knee moment of inertia. These collapse diagrams show the probable passing of the frames drift from the particular drift (threshold of the limit state) mentioned in the diagram for the future earthquakes.

Dynamic analysis results of 4 and 8 story frames with constant knee moment of inertia and variable length

Here, for the sake of brevity, only the results of the incremental dynamic analysis of the 4 and 8 story frames are presented.

As it can be seen in the average diagrams of Figs. 12 and 13, increasing the knee length in these frames has caused the average response of the frame to increase, i.e., in a lower earthquake intensity, the frame experiences a larger drift and the average diagrams of the frames with larger knee length is below that of the frames with the smaller knee length. This means that the strength of the

frames with a larger knee length is lower than that of the frames with smaller knee length to better understand the subject, for an increase in the earthquake's intensity, the collapse diagrams in 2, 3 and 4 % are presented in Figs. 14 and 15.

As it can be observed in the collapse diagrams for the aforementioned drift, the diagram of the larger knee length is above the diagram of the smaller knee length. i.e., the frame with the larger knee length is more lightly to fail than the frame with the smaller knee length.

Conclusion

Results of the incremental dynamic analysis show that the response of a particular structure different accelerograms varies according to the specifications of the structure and the earthquake such as duration of the earthquake, frequency, amplitude, etc., at times these differences are so significant that the responses of multiple accelerograms cannot be compared. The diagrams represent the efficient behavior of the knee bracings as compared to the diagonal bracings. In the diagonal-braced frames, the drifts of the building greatly increase in one or two stories due to the severe buckling of the bracings under seismic forces, but in the knee-braced frames, before buckling of the bracing, the knee members will yield, start to absorb the energy, and prevent buckling of the bracing.

The drift of the stories under seismic forces is higher in the knee members with great length and weak moment of inertia, and, as a result, when the length of the knee is greater, the use of lower moments of inertia for the knee is not recommended.

The summary of the average IDA diagram for frames with smaller knee length, in all conditions (4 and 8 story frame) is higher than the average diagram of the frames with larger knee lengths, i.e., the frames with a large knee length experience a larger drift as the intensity of the earthquake increases. Thus, increasing the knee length causes the earthquake resistance of the frames to diminish.

The summary of the average IDA diagrams for frames with larger knee moment of inertia, in all conditions is higher than the average diagrams of frames with smaller knee moment of inertia, i.e., the frames with low knee moment of inertia experience a larger drift as the intensity of the earthquake increases. Thus decreasing the moment of inertia causes the earthquake resistance of the frame to diminish.

Collapse diagram of the frame with the larger knee length was above the collapse diagram of the frame with the smaller knee length, i.e., the frame with the larger knee length more probable to collapse than the frame with the smaller knee length.

At small knee lengths, the use of the strong moments of inertia leads to the buckling of the brace; therefore, using the knee element with a small length and a high moment of inertia is not recommended. However, even in the worst conditions, the behavior of this system is more efficient than that of the diagonal system.

One of the most influential parameters in the collapse diagrams is the selected accelerograms to analyze the structure. As it was seen in the IDA diagrams, the structure shows hardening behavior to some accelerograms and softening behavior to some others and thus it possible for the accelerogram to be selected in such away that all or most of them have stiffening behavior which makes the resulted collapse diagram to show a low collapse probability. In contrast, the accelerograms can be selected in such a way that the diagrams display softening behavior and the resulted collapse diagram show a high collapse probability. Thus the selection of the accelerograms affects the collapse diagrams and the acquired collapse diagrams here in are based on the accelerograms used in this study.

References

Balendra T, Sam MT (1990) Diagonal brace with ductile knee anchor for a seismic steel frame. Earthq Eng Struct Dyn J 19(6):847–858

Balendra T, Sam MT (1991a) Design of earthquake resistance steel frame with knee bracing. Earthq Eng And Struct Dyn J 19(6):847–858

Balendra T, Sam MT (1991b) Preliminary studies into the behavior of knee braced frames subject to seismic loading. Earthq Eng And Struct Dyn J 13:67–74

Balendra T, Sam MT (1994) Ductile knee braced frame with shear yielding knee for seismic resistance structures. Earthq Eng Struct Dyn J 19(6):847–858

BHRC Iranian code of practice for seismic resistance design of buildings: standard No. 2800. 3rd ed. (2005) Building and Housing Research Center

Chopra AK (1998) Dynamic Of Structures. The theory and application Earthq Eng, Prentice Hall New Delhi

Hjelmstad KD, Popov EP (1983) Seismic behavior of active beam links in eccentrically braced frames. Earthq Eng Research Center Report No UCB/EERC-83/15 Berkeley (CA) University of California

Inouel K, Suita K, Takeuchi I, Chusilp P, Nakashima M, Zhou F (2006) Seismic-resistant weld-free steel frame buildings with mechanical joints and hysteretic dampers. J of Struct Eng 132(6):864–872

Jinkoo K (2005) Response modification factor of chevron braced frames. Earthq Eng And Struct Dyn J 27:285–300

Lee K, Bruneau M (2005) Energy dissipation of compression members in concentrically braced frames. J Struct Eng 131(4): 552–559

McKenna F, Fenves GL, Scott MH (2007) Open system for earthquake engineering simulation. Pacific Earthquake Engineering Research Center, University of California, Berkeley. Available from http://opensees.berkeley.edu

Mofid M, Khosravi P (2000) Non-linear analysis of disposable knee bracing. Comput Struct J 75:65–72

Mofid M, Lotfollahi M (2006) On the characteristics of new ductile knee bracing system. J Const Steel Res 62:271–281

Ochao A (1986) Disposable knee bracing improvement in seismic design of steel frames. J Struct Eng ASCE 7(112):1544–1552

Seismic provisions for structural steel buildings (2005) American Institute of Steel Construction, AISC

Sutat L, Bunyarit S, Jarun S, Pennung W (2011) Seismic design and behavior of ductile knee-braced moment frames. J Struc Eng 137(5)

William M, Denis C (2004) Seismic design and analysis of a knee braced frame building. Earthq Eng And Struct Dyn J 23(6):138–258

Zahrai M, Jalali M (2014) Experimental and analytical investigations on seismic behavior of ductile steel knee braced frames. Steel and Composite struct 16(1) 1–21

11

Structural performance of notch damaged steel beams repaired with composite materials

Boshra El-Taly[1]

Abstract An experimental program and an analytical model using ANSYS program were employed to estimate the structural performance of repaired damaged steel beams using fiber reinforced polymer (FRP) composite materials. The beams were artificially notched in the tension flanges at mid-spans and retrofitted by FRP flexible sheets on the tension flanges and the sheets were extended to cover parts of the beams webs with different heights. Eleven box steel beams, including one intact beam, one notch damaged beam and nine notches damaged beam and retrofitted with composite materials, were tested in two-point loading up to failure. The parameters considered were the FRP type (GFRP and CFRP) and number of layers. The results indicated that bonding CFRP sheets to both of the tension steel flange and part of the webs, instead of the tension flange only, enhances the ultimate load of the retrofitted beams, avoids the occurrence of the debonding and increases the beam ductility. Also the numerical models give acceptable results in comparison with the experimental results.

Keywords Box steel beam · Composite materials · Retrofit methods · Finite element analysis · Non-linear analysis

Introduction

FRP composite materials were initially used as strengthening materials for reinforced concrete (RC) flexural components then their applications have been expanded to wood, masonry, steel and concrete–steel composite structures. The composite materials may be glass or carbon-FRP materials (GFRP or CFRP) and they are used in the form of flexible sheets or rigid plates. They have light weight, high tension strength and high resistance to corrosion. FRP is fixed in the structure by adhesive (resin) material. The bond at the interface between the structure and FRP controls in the transferring force between them. The researches effort in the field of retrofitting steel I beam by bonding CFRP plate into the beam lower flange showed that the presence of the CFRP plate can help to increase the ultimate strength and post-elastic stiffness of typical I-steel beams (especially when a high modulus CFRP is used). Also various failure modes occurred for such FRP-plated steel beams; plate end debonding in an FRP-plated steel beam is the basic mode of failure. This because high interfacial shear stresses and peeling stresses localize in the vicinity of the plate end. Plate end debonding can be delayed by increasing the bonded length (Schnerch et al. 2007; Tavakkolizadeh and Saadatmanesh 2003; Photiou et al. 2006a, b; Michael et al. 2005; Gu et al. 2012; Jesus et al. 2012; Buyukozturk and Hearing 1999; Shaat and Fam 2006, 2009; Teng et al. 2012; Salama and Abd-El-Meguid 2010; Yu et al. 2011; Hmidan et al. 2013; Ochi et al. 2011; Kalavagunta et al. 2014).

Various researches examined the feasibility and effectiveness of rehabilitation of deteriorated steel structures using FRP materials. Deterioration in steel structures occurs due to excessive service load, fatigue damage or environmental contribution (corrosion). The researches

✉ Boshra El-Taly
boushra_eltaly@yahoo.com

[1] Lecturer in Civil Engineering Department, Faculty of Engineering, Minoufia University, Shibin Al Kawm, Egypt

efforts in this field generally focused on repairing naturally deteriorated steel structure elements or repairing artificially damaged steel structure elements. In 1996, two I-girders were removed from an old and deteriorated bridge. The two girders had uniform corrosion along their length and this corrosion mostly concentrated within the tension flange and the webs of the girders were not severely corroded. This corrosion caused approximately, a 40 % loss of the tension flange. The two beams were repaired along the entire length of the girders using a single layer of CFRP strip with 6.4 mm thick then they were tested. The results indicated that the elastic stiffness of the first and second girders increased by 10 and 37 %, respectively. The ultimate capacities of the first and second girders were also increased by 17 and 25 %, respectively (Gillespie et al. 1996).

On the other hand various researches studied the structural performance of repairing artificially damaged composite or non-composite steel beam using FRP element. Section loss due to corrosion has been simulated artificially by cutting part of the flange or the web, or machining the tension flange to a reduced thickness throughout the entire span. Fatigue cracks have been simulated by introducing partial or complete saw cut in the steel flange thickness. Another method of introducing an artificial damage is by loading the steel girder beyond yielding and then unloading. Liu et al. (2001) studied the effect of repairing damaged non-composite I-beams using HM-CFRP plate. The tension flange of the damaged beams was completely cut. Their results indicated that the repaired specimens did not recover the total strength of the control intact specimen. The failure mode of specimen with full length repair was due to a gradual debonding of the CFRP laminate which initiated at mid-span and extended to the end as the load increased. Tavakkolizadeh and Saadatmanesh (2003) and (Al-Saidy et al. 2004) examined the effect of repairing partial cutting of the tension flange of composite beams using SM-CFRP sheets and HM-CFRP plates, respectively. Tavakkolizadeh and Saadatmanesh (2003) indicated that the repaired specimens did not recover the total strength of the intact specimen. Also their results showed that the girder having 25 % loss in tension flange failed by rupture of CFRP. On the other hand the girder having 50 % loss in tension flange failed by crushing of the concrete slab, followed by a limited debonding of the CFRP laminate at mid-span. In the case of the girder having 100 % loss in tension flange complete debonding of CFRP laminate occurred. Al-Saidy et al. (2004) showed that repairing the girders using HM-CFRP plates was able to fully restore the strength of the original undamaged girders. The CFRP debonding mode of failure was not observed in their study and only crushing of the concrete slab or rupture of the CFRP plates was reported. Shaat (2007) studied the effect

of repairing the damage composite steel beams by bonding SM-CFRP or HM-CFRP sheets on the bottom side of the tension steel flange and on both the bottom and top sides of the tension flange. The steel tension flanges were completely cut at mid-span to represent a severe section loss in bridge girders. His results indicated that bonding SM-CFRP sheets to both sides of the steel flange enhances flexural stiffness of the repaired girders, but has no effect on flexural strength.

Eleven box steel beams (one control intact beam, one artificially damaged beam and nine artificially damaged beams and repaired with FRP composite material) were tested in four-point bending configuration. The tension flange is saw-cut at mid-span to simulate section loss due to a fatigue crack or a localized severe corrosion. FRP material is then adhesively bonded to the tension flange or to the tension flange and parts of the webs in order to recover and possibly exceed the original strength. Analytical models using ANSYS program were employed, verified, and used in a parametric study. The current results indicate that the FE simulations give acceptable results in comparison with the experimental results.

Experimental program

Eleven steel box beams were tested in two-point loading. All beams are of length 1000 mm and they have box cross section with 50×50 mm and with 3 mm thickness. A schematic diagram of the typical beam (CB) and the locations of strain and dial gages are shown in Fig. 1. The specimens were categorized into four groups. The first group (Set 1) is the control specimens and it includes the control intact beam (CB) and the damaged beam (DB). The second group (Set 2) contains four artificially damaged beams and then they were repaired with FRP composite material. In this group, the damaged beams were repaired by externally bonded the composite material on the tension flange only (repaired method Case # 1) as shown in Fig. 2. The type of FRP; glass fiber or carbon fiber and the number of layers are the parameters. In the third group (Set 3), the specimens were retrofitted by attaching the composite material in the tension-side flanges and the FRP flexible sheets were wrapped around the beams web to cover parts of them with H height as shown in Fig. 2. The fourth group (Set 4) included one retrofitted beam. This beam was repaired by attaching two layers of CFRP sheets on the lower tension flange and applied four GFRP strips with 20 mm width at the CFRP ends. These strips were wrapped around the whole section of the beam as shown in Fig. 3. In all the damaged specimens, the tension flange is saw-cut at mid-span to simulate section loss due to a fatigue crack or a localized severe corrosion. The width and depth of the

Fig. 1 Beam details, strain and dial gage locations and notch details

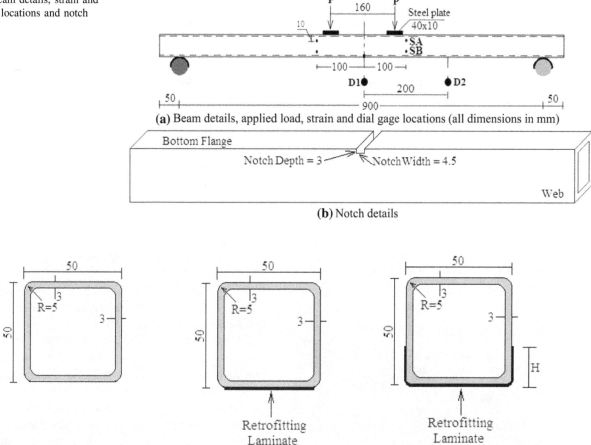

(a) Beam details, applied load, strain and dial gage locations (all dimensions in mm)

(b) Notch details

Fig. 2 Cross section details and method of retrofitting (all dimensions in mm)

Typical cross section   Repairing method (Case # 1)   Repairing method (Case # 2)

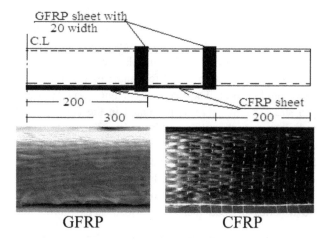

GFRP   CFRP

Fig. 3 Details of RB9 specimen

notch were controlled to be 4.5 and 3 mm, respectively in all the damaged specimens. Details of the test specimens are summarized in Table 1.

The glass fiber used is a unidirectional woven glass fiber fabric with 0.172 mm thickness and the CFR is SikaWrap®-230 C that is a unidirectional woven carbon fiber fabric for the dry application process. Its thickness is 0.131 mm. Sikadur®-330 supplied by Sika Company was used to get sufficient bonding between steel beam and both of carbon fibre and glass fiber. It is a two parts (A and B). Part A is a resin (white) and the second part is a hardener (gray). The mixing ratio was 4:1 (A: B) by weight. Before applying the strengthened materials, the surface was clean from any dust by manual sandblasting. Table 2 shows the material properties of the CFRP, GFRP, steel and adhesive material.

The specimens were simply supported with a span of 900 mm between the two supports and tested under two-point bending static loading using flexural testing machine of 100 kN capacity. The spacing between the two concentrated point loads was 160 mm (see Fig. 1). The beams were tested under an increasing load up to failure at 1 kN a constant loading rate. The behavior of beams was monitored by measuring the deflection at mid-span and at 200 mm distance from the mid-span using two dial gauges with an accuracy of 0.01 mm. A set of four DEMEC strain gauges were fixed in one side of the specimen to allow

Table 1 Specimens details

Group I.D.	Specimens I.D.	Beam designation	Retrofitting configuration
Set 1	CB	Intact	None
	DB	Damaged	None
Set 2	RB1	Repaired method Case # 1	2 layer of GFRP, 600 mm long
	RB2		1 layer of CFRP, 600 mm long
	RB3		1 layer of CFRP, 600 mm long + 1 layer of CFRP, 400 mm long
	RB4		2 layer of GFRP, 600 mm long + 2 layer of GFRP, 400 mm long
Set 3	RB5	Repaired method Case # 2	2 layer of GFRP, 600 mm long (web height (H) = 10 mm)
	RB6		1 layer of CFRP, 600 mm long (web height (H) = 5 mm)
	REP7		2 layer of GFRP, 600 mm long + 2 layer of GFRP, 400 mm long (web height (H) = 15 mm)
	RB8		1 layer of CFRP, 600 mm long + 1 layer of CFRP, 400 mm long (web height (H) = 15 mm)
Set 4	RB9	Repaired method Case # 1	1 layer of CFRP, 600 mm long + 1 layer of CFRP, 400 mm long + GFRP strips at the ends of CFRP

Fig. 4 Specimen test

measuring the strain versus load during the test. Typical test set-up and instrumentation is shown in Fig. 4.

FE simulation

Three dimensional models were employed to simulate the tested steel beams numerically up to failure using the general purpose Finite Element (FE) analysis program, ANSYS. Geometric and material nonlinearities were both considered into account in simulating the flexural behaviors of the retrofitted and non-retrofitted beams. Solid 185 elements (refer to ANSYS (Narmashiri and Jumaat 2011) and Narmashiri and Jumaat (ANSYS 2009)) were used to simulate the steel beams and the strengthening composite materials and the adhesive material. Each element is defined by eight nodes. Each node has three degrees of freedom at each node: translations in the nodal x, y, and z directions. The element has plasticity, creep, swelling, stress stiffening, large deflection, and large strain capabilities. The modeling included beams having a hinged support on one side and a roller support at the other side, with a span length of 900 mm as per the experimental setup. The FE simulation of the control beam as example is presented in Fig. 5. The material nonlinearity was represented by multi-linear kinematic hardening constants (MKIN). It assumes that the total stress range is equal to twice the yield stresses, so that Bauschinger effect is included. MKIN may be used for materials that obey von Mises yield criteria. The material behavior was described by a stress–strain curve as presented in the experimental work. It starts at the origin and it is with positive stress and strain values. The initial slope of the curve represents the elastic modulus of the material. In the current analysis, load-control technique is used. In this technique, total load is applied to a finite element model. The load is divided into a series of load increments (load steps) during the

Table 2 Material properties of steel and composite materials

	Density g/cm³	E-Modulus (Gpa)	Tensile strength (Mpa)		Strain	
			Yield	Ultimate	Yield	Ultimate
Steel	7.8	200	250	370	0.00125	0.0135
CFRP	1.76	238	4300	–	0.018	–
GFRP	2.56	18.3	381	–	0.0208	–
Adhesive		3.8	4500	–	0.009	

Fig. 5 Geometry of the FE model

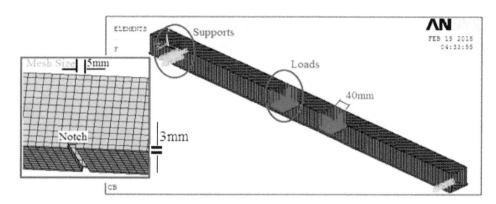

Table 3 Test results

Specimens I.D.	At the end of linear edge			At the peak load					Failure mode
	P_L (kN)	D_{1L} (mm)	D_{2L} (mm)	P_u (kN)	D_{1u} (mm)	D_{2u} (mm)	Strain (µ£)		
							Com.	Ten.	
CB	15.3	8.25	5.62	17.8	30.15	24.42	6900	5120	Ductile
DB	8.3	4.08	3.13	11.8	16.08	11.22	800	760	
RB1	8.3	3.77	2.7	12.3	10.89	6.88	3700	1380	Rupture of the laminates
RB2	9.3	4.19	3.61	14.5	14.59	9.21	1440	1140	
RB3	13	6.26	4.76	18.7	27.15	18.18	1720	2320	Debonding
RB4	8.4	4.04	3.1	14.4	24.24	12.6	1840	1520	Rupture of the laminates
RB5	12	5.32	3.81	14.6	16.67	11.42	1380	1620	
RB6	11	5	3.82	17	19.3	13.25	3880	4480	
RB7	13	5.06	3.6	16.4	17.55	11.16	2360	1040	
RB8	14	6.04	4.65	22.5	32.04	23.37	12780	3240	
RB9	14.5	8.47	5.25	19.5	24.34	18.05	6660	1680	Debonding

analysis. ANSYS program uses Newton–Raphson method for updating the model stiffness [refer to ANSYS Tavakkolizadeh and Saadatmanesh (2003) and Kadhim Shaat and Fam (2006)]. Finally, initial imperfections or residual stresses caused by the manufacturing process were not included in the modeling.

Results and discussion

The experimental test results of all the steel beams are summarized in Table 3. In the current section, the effect of cutting the tension flange at mid-span is presented and discussed. Figure 6 shows the load versus deflection at points D1 and D2 from the experimental work and FE simulation of beam CB and beam DB. Figure 7 shows the load versus longitudinal strains at the measured locations (SA and SB) for the two specimens. The failure mode as observed from the experimental work and as obtained from the numerical analysis for the two beams are showed in Fig. 8. From these figures, it can be concluded that the finite element simulation

gives acceptable results in comparison with the experimental results. From Fig. 6, it can be concluded that the relation between experimental load and the deflection at the mid span of beam CB is linear up to 15.3 kN total load and 8.25 mm corresponding mid-span deflection. After that the plasticity took place and growth in the bottom flange, the top flange and the two webs at the loading positions and their surrounding area, making load–deflection curve nonlinear up to an applied load equals 17.8 kN and the corresponding deflection at mid-span equals 15.5 mm, after that the deflection increased without significant change in the load. Also this figure shows that the strength of beam DB has been severely degraded as a result of notching the lower steel flange at mid-span. Table 3 shows 34 % reduction in the ultimate load. Figure 8 shows that the deformed shape at the ultimate load as observed from the experimental work and as obtained from the numerical analysis for beam CB is ductile failure mode. For the damaged beam, significant yielding was observed and at the end of the test, the measured crack width was 7.2 mm. Figure 9 shows numerical normal stresses distribution for the two beams at the ultimate load. This

Fig. 6 Load-deflection curves of non-retrofitted specimens; *left* at point D1 and *right* at point D2

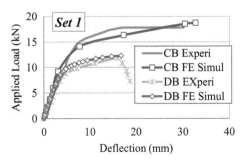

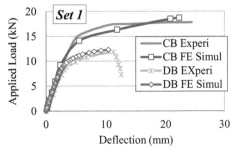

Fig. 7 Experimental and FE simulation total load-strain curves of non-retrofitted specimens

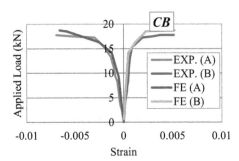

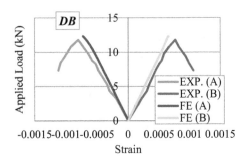

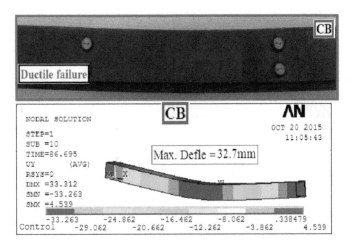

Fig. 8 Failure modes of non-retrofitted specimens

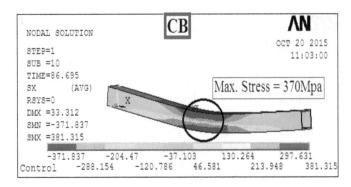

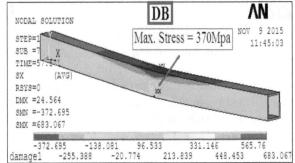

Fig. 9 Numerical normal stresses distribution at the ultimate load of non-retrofitted specimens

Fig. 10 Load-deflection curves of retrofitted specimens (Set 2); *left* at D1 and *right* at D2 point

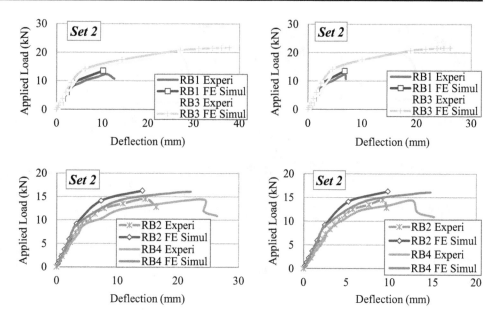

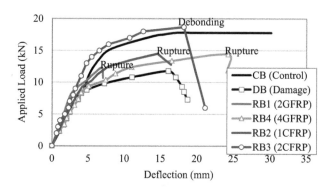

Fig. 11 Experimental total load–deflection curves of retrofitted specimens (Set 2)

figure shows that the region around the cut edge (the notch) of the flange in notch damaged beams yields much earlier than farther regions along the web, due to stress concentration.

The results of repaired beams with the composite materials bonded on the lower side of the tension steel flange only (Set 2) in terms of load–deflection curves, load-strain curves, failure modes and normal stresses distributions are presented in Figs. 10, 11, 12, 13. Figure 10 shows the comparison between the experimental and numerical results in terms of load–deflection curves for retrofitted specimens (Set 2). This figure shows that the numerical and the experimental results are close; the difference between

Fig. 12 Load-strain curves of retrofitted specimens (Set 2); Load-strain curves of retrofitted specimens (Set 2)

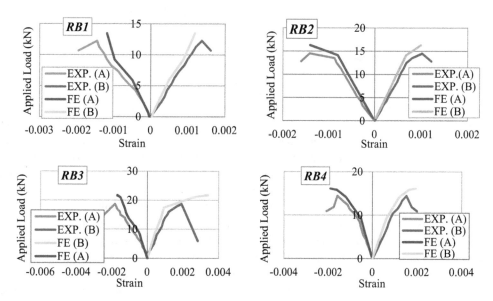

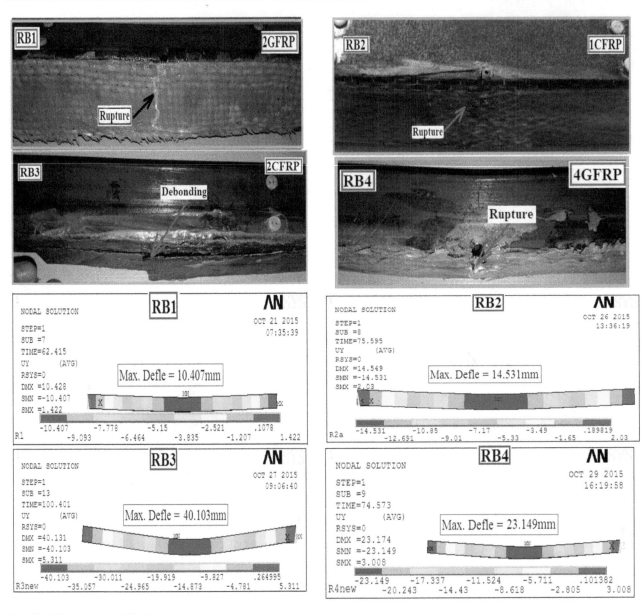

Fig. 13 Failure modes of Set 2 specimens

the results not exceed than 15 %. Figure 11 shows the responses of specimens Set 2, compared to beams CB and DB. The figure shows that beam RB3 reached flexural stiffness values higher than the intact beam, which is attributed to the high value of elastic modulus of CFRP. Table 3 shows gains in ultimate load of 5.1 %. This beam failed at 17.8 kN ultimate load and 27.15 and 18.18 mm corresponding deflections at points D1 and D2 due to the occurrence of the debonding CFRP as shown in Fig. 13. Also Fig. 11 indicates that the retrofitted beams (RB1, RB2 and RB4) give higher responses than the notch damaged beam but they not achieve the full stiffness and ductility of beam CB and their ultimate load was 30.9, 18.54 and

19.10 % lower than the ultimate load of the intact beam, as indicated in Table 3. Also these results showed that increasing the number of FRP layers increases the ultimate load. Figure 12 shows the load versus strains at two different locations (SA and SB) of Set 2 specimens. From these figures, it can be clearly seen that the numerical model gives a good outcome compared to the experimental tests results. The failure mode of beams RB1, RB2 and RB4 occurred by rupture of the FRP sheets at mid-span, as shown in Fig. 13 due to reaching the maximum strength of FRP (see Fig. 14).

The structural performance of the retrofitted specimens with the composite materials bonded on the lower side of the

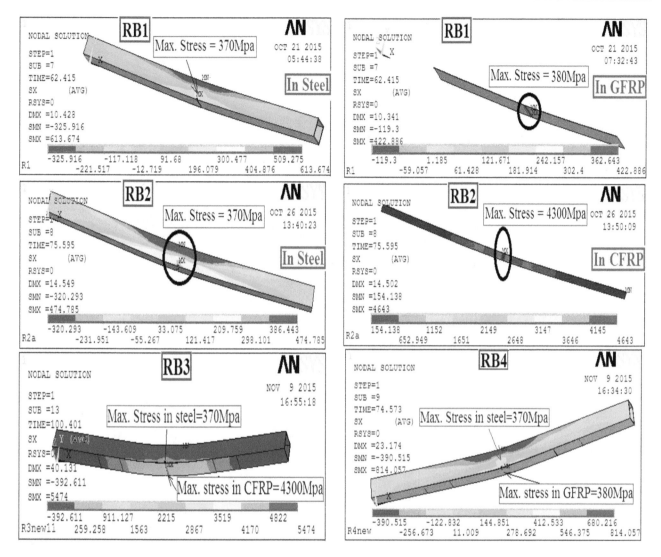

Fig. 14 Numerical normal stresses at the ultimate load of (Set 2) retrofitted specimens

tension steel flange and on a part of the two webs with height (*h*) (Set 3) are presented and discussed in this section. Figure 15 shows the experimental and numerical load–deflection curves at points D1 and D2 and Fig. 16 presents the comparison between experimental and numerical results of beams Set 3 in term of load verses strain. These figures showed that the FE simulation gives good results compared to the experimental results. Figure 17 shows the experimental load–deflection curves of specimens Set 3 (RB5, RB6, RB7 and RB8), compared to the reference control intact beam and the damaged beam. This figure and Table 3 indicate that beam RB8 achieves ultimate load values higher than beam CB by about 26.4 %. Also beam RB8 has an increase in ductility in comparison with beam CB. This beam failed at 22.5 kN ultimate load and 32.04

and 23.37 mm corresponding deflections at points D1 and D2 due to the occurrence of the rupture CFRP at the mid-span as shown in Fig. 18. The failure mode of all beams Set 3 occurred by rupture of the FRP sheets at mid-span, as shown in Fig. 18. Figure 19 presents the numerical normal stresses of retrofitted beams Set 3.

The results of the specimen RB9 are presented in Figs. 20, 21, 22, 23. This specimen was repaired with two layers of CFRP and four externally bonded GFRP strips. These strips were wrapped around the whole section of the beam at the ends of the CFRP layers. Figure 21 indicates that the repaired specimen RP9 achieves ultimate load value higher than beam CB by about 9.55 %. Figure 22 shows that the end debonding is avoid in this beam and the debonding initially appeared at the mid-span.

Fig. 15 Load-deflection curves of retrofitted specimens (Set #3); *left* at D1 and *right* at D2 point

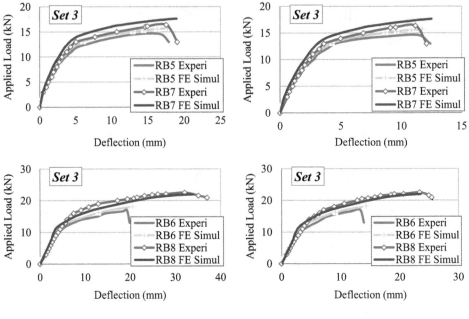

Fig. 16 Experimental and FE simulation load-strain curves of retrofitted specimens (Set 3)

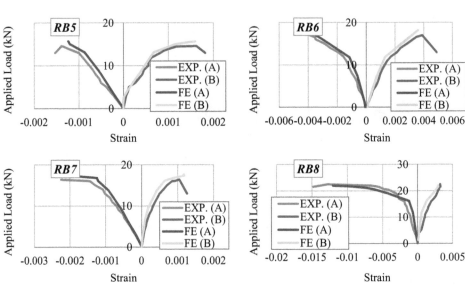

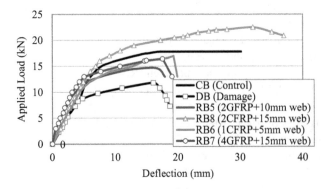

Fig. 17 Experimental total load–deflection curves of retrofitted specimens (Set 3)

Conclusions

The main goal of this study was to study repairing of damaged box steel beams to recover their original capacities. Experimental and numerical investigations were carried out for testing eleven box steel beams (control intact beam, notch damaged beam and nine retrofitted beams) in two-point bending. The nine beams were artificially damaged and have been repaired using FRP sheets. The parameters considered were the effect of FRP type (GFRP and CFRP), number of layers and the effect of wrapping the composite material to cover a part of the webs. Based on these studies, the following conclusions are drawn:

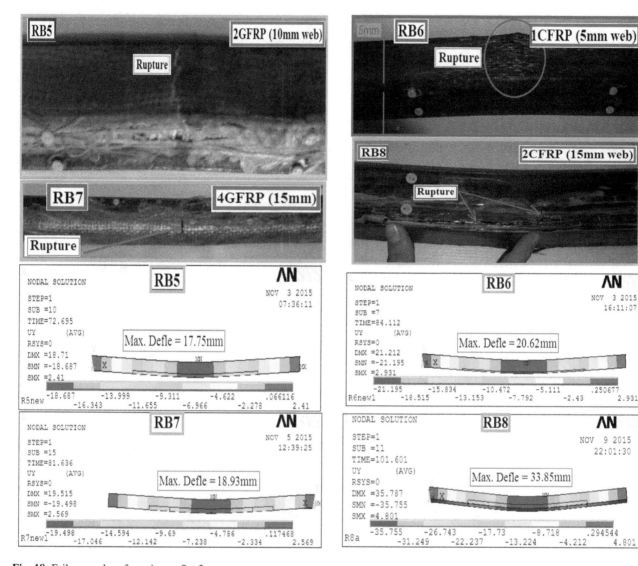

Fig. 18 Failure modes of specimens Set 3

1. The FE simulations give acceptable results in comparison with the experimental results.
2. The flexural strength of beam DB tested has been reduced by 34 % as a result of notching of the tension flange at the mid-span.
3. The failure mode of the intact beam is ductile mode failure. Also the region around the cut edge (the notch) of the flange in notch damaged beam yields earlier than farther regions along the web because of stress concentration.
4. The retrofitted beams with CFRP reached ultimate load values higher than the retrofitted beams with GFRP.
5. RB3 repaired beam reached ultimate load value higher than beam CB (an increase in ultimate load of 5.1 %). On the other hand specimen RB8 restored the ultimate

strength of beam CB and even exceeded them by 26.4 % and it has an increase in ductility in comparison with beam CB.

6. The specimen repaired with two layers of CFRP bonded on the lower flange and four GFRP strips wrapped around the whole section of the beam at the ends of the CFRP layers achieves an increase in the ultimate load than beam CB by about 9.55 %.

7. The girder repaired using one layer CFRP sheets failed by rupture of the sheets and the beam repaired using two layers CFRP sheets attached with the tension flange only failed by debonding at the end of the sheets. This debonding is hidden by wrapping the sheet on the web in beam RB8 and by wrapping strips of GFRP at the CFRP sheet ends.

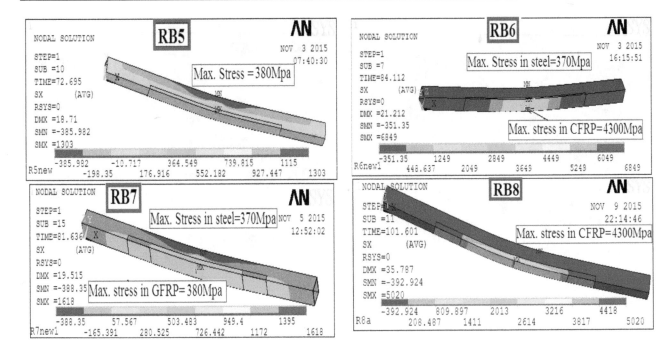

Fig. 19 Numerical normal stresses at the ultimate load of retrofitted specimens Set 3

Fig. 20 Load-deflection curves of retrofitted specimens Set 4

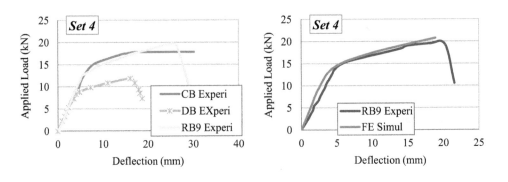

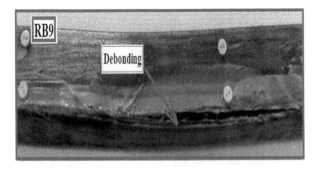

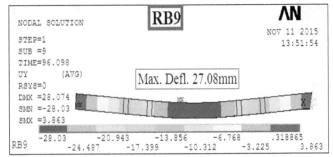

Fig. 21 Failure modes of specimens Set 4

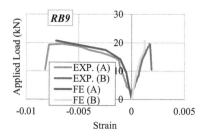

Fig. 22 Load-strain curves of beam RB9

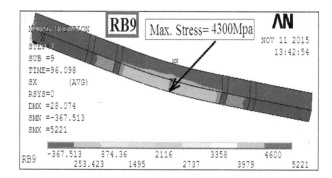

Fig. 23 Normal stresses at the ultimate load of beam RB9

8. Experimental and numerical results indicate the effectiveness of CFRP, as a candidate retrofitting material, for damaged steel structures.

References

Al-Saidy AH, Klaiber FW, Wipf TJ (2004) Repair of steel composite beams with carbon fiber-reinforced polymer plates. J Compos Constr ASCE 2(2):163–172

ANSYS (2009) ANSYS Help. Release 12.0, Copyright

Buyukozturk O, Hearing B (1999) Failure behavior of pre-cracked concrete beams retrofitted with FRP. J Compos Constr (ASCE) 3(2):138–144

Gillespie JW, Mertz DR, Edberg WM, Ammar N, Kasai K, Hodgson IC (1996) Rehabilitation of steel bridge girders through application of composite materials. In: 28th International SAMPE Technical Conference. November 4–7, pp 1249–1257

Gu X, Peng B, Chen G, Li X, Ouyang Y (2012a) Rapid strengthening of masonry structures cracked in earthquakes using fiber composite materials. J Compos Constr (ASCE) 5(16):590–603

Hmidan A, Kim Y, Yazdani S (2013) Crack-dependent response of steel elements strengthened with cfrp sheets. Constr Build Mater 49:110–120

Jesus AMP, Pinto JMT, Morais JJL (2012) Analysis of solid wood beams strengthened with cfrp laminates of distinct lengths. Constr Build Mater 35:817–828

Kadhim M (2012) Effect of CFRP Plate length strengthening continuous steel beam. J Constr Build Mater 28:648–652

Kalavagunta S, Naganathan S, Mustapha K (2014) Axially loaded steel columns strengthened with CFRP. Jordan J Civil Eng 8(1):58–69

Liu X, Silva PF, Nanni A (2001) Rehabilitation of steel bridge members with FRP composite materials. In: Proceedings of the international conference on composites in construction, October 10–12, Porto, Portugal, pp 613–617

Michael J, Chajes MJ, Chacon AP, Swinehart MW, Richardson D, Wenczel R, Liu W (2005) Applications of advanced composites to steel bridges. Technical Report, Delaware Center for Transportation, University of Delaware, Newark

Narmashiri K, Jumaat MZ (2011) Reinforced Steel I-Beams: a Comparison between 2D and 3D simulation. J Sim Model Pract Theory 19:564–585

Ochi N, Matsumura M, Hisabe N (2011) Experimental study on strengthening effect of high modulus cfrp strips with different adhesive length installed onto the lower flange plate of i shaped steel girder. J Proc Eng 14:506–512

Photiou NK, Hollaway LC, Chryssanthopoulos MK (2006a) Strengthening of an artificial degraded steel beam utilizing a carbon/glass composite system. Constr Build Mater 20(1–2):11–21

Photiou NK, Hollaway LC, Chryssanthopoulos MK (2006b) Selection of Carbon-Fiber-Reinforced Polymer Systems for Steelwork Upgrading. J Mater Civ Eng ASCE 5(18):641–649

Salama T, Abd-El-Meguid A (2010) Strengthening steel bridge girders using CFRP. Technical Report, University Transportation Center for Alabama (UTCA), The University of Alabama, Birmingham, No. 06217, pp 1–184

Schnerch D, Dawood M, Rizkalla S, Sumner E (2007) Proposed design guidelines for strengthening of steel bridges with FRP materials. Constr Build Mater 2:1001–1010

Shaat A (2007) Structural behavior of steel columns and steel-concrete composite girders retrofitted using CFRP. Ph. D, Thesis, Queen's University, Kingston, Ontario

Shaat A, Fam A (2006) Axial loading tests on short and long hollow structural steel columns retrofitted using carbon fibre reinforced polymers. Can J Civ Eng 4(33):458–470

Shaat A, Fam A (2009) Slender steel columns strengthened using high- modulus cfrp plates for buckling control. J Compos Constr (ASCE) 1(13):2–12

Tavakkolizadeh M, Saadatmanesh H (2003a) Strengthening of steel-concrete composite girders using carbon fiber reinforced polymers sheets. J Struct Eng (ASCE) 1(129):30–40

Tavakkolizadeh M, Saadatmanesh H (2003b) Repair of damaged steel-concrete composite girders using carbon fiber reinforced polymers sheets. J Compos Constr ASCE 4(7):311–322

Teng JG, Yu T, Fernando D (2012) Strengthening of steel structures with fiber-reinforced polymer composites. J Construct Steel Res 78:131–143

Yu Y, Chiew SP, Lee CK (2011) Bond failure of steel beams strengthened with FRP laminates—part 2: verification. J Compos Part B 42:1122–1134

12

The influence of coupled horizontal–vertical ground excitations on the collapse margins of modern RC-MRFs

Ehsan Noroozinejad Farsangi[1] · Abbas Ali Tasnimi[2]

Abstract With the increasing interest in vertical ground motions, the focus of this study is to investigate the effect of concurrent horizontal–vertical excitations on the seismic response and collapse fragilities of RC buildings designed according to modern seismic codes and located near active faults. It must be stressed that only mid- to high-rise buildings are of significant concern in the context of this research. The considered structures are categorized as intermediate and special RC-MRFs and have been remodeled using distributed and lumped plasticity computational approaches in nonlinear simulation platforms, so that the utilized NL models can simulate all possible modes of deterioration. For better comparison, not only was the combined vertical and horizontal motion applied, but also a single horizontal component was considered for direct evaluation of the effect of the vertical ground motions (VGMs). At the member level, axial force variation and shear failure as the most critical brittle failure mechanisms were studied, while on the global level, adjusted collapse margin ratios (ACMRs) and mean annual frequency of collapse ($\lambda_{\text{Collapse}}$) using a new vector-valued intensity measure were investigated. Findings from the study indicate that VGMs have significant effects on both local and global structural performance and cannot be neglected.

Keywords Vertical excitation · Vector IM · Adjusted collapse margin ratios (ACMRs) · Mean annual frequency (MAF) of collapse · Seismic fragility · Nonlinear (NL) models

List of symbols

$d_{\text{max},i}$	Current deformation that defines the end of the reload cycle for deformation demand
F_i^+ and F_i^-	Deteriorated yield strength after and before excursion i, respectively
$F_{\text{ref}}^{+/-}$	Intersection of the vertical axis with the projection of the post-capping branch
F_y	Yield strength
k_1 and k_u	Constants specifying the lower and upper bounds in the vector IM
K_0	Element stiffness at Δ_{cr}
K_1	Element stiffness at Δ_y
K_2	Element stiffness at Δ_m
K_{deg}	Degrading slope of the shear spring based on the limit-state material
K_e	Elastic (initial) stiffness of the element
K_{rel}	Reloading stiffness of the element
K_s	Slope of the hardening branch
$K_{\text{deg}}^{\text{t}}$	Degrading slope for the total response in OpenSees model
$K_{u,i}$ and $K_{u,i-1}$	Deteriorated unloading stiffness after and before excursion i, respectively
$K_{\text{unloading}}$	Unloading slope of the rotational spring in the OpenSees model
T_1	Dominant period of vibration for a specific structure
T_h	Horizontal period of vibration for a specific structure
T_{low} and T_{upp}	Lower and the upper periods of the elastic spectrum
T_v	Vertical period of vibration for a specific structure

✉ Abbas Ali Tasnimi
tasnimi@modares.ac.ir

[1] International Institute of Earthquake Engineering and Seismology (IIEES), Tehran, Iran

[2] Tarbiat Modares University, Tehran, Iran

V_{cr}	Shear force corresponding to displacement which causes concrete cracking
V_m	Shear force corresponding to the maximum displacement
V_y	Shear force corresponding to the displacement which causes steel yielding
β_c, $\beta_{D|IM}$ and β_M	Uncertainties in capacity, demand and modeling
β_{TOT}	Total uncertainty
δ_{Ci}	Cap deformation at the i-th cycle
$\delta_{t,i}^{+/-}$	Target displacement for each loading direction at the i-th cycle
Δ	Total deformation
Δ_{cr}	Deformation at cracking
Δ_f	Flexural deformation
Δ_m	Maximum displacement
Δ_s	Shear deformation
Δ_y	Deformation at yielding
$\Phi(.)$	Standard normal cumulative distribution function
ρ_t	Transverse reinforcement ratio in beams and columns
$\lambda_{Collapse}$	Mean annual frequency of collapse
$\lambda_{IM}(x)$	Mean annual frequency of the ground motion intensity exceeding x
μ_T	Period-based ductility
χ_c and $\chi_{D|IM}$	Natural logarithm of the median capacity and demand of the structural system

Introduction

Earthquakes in the past have indicated that enormous damage to the building structures and human casualties will result, in the case of severe seismic events. Hence, vulnerability assessment and seismic loss estimation are the primary concerns for regulatory agencies and civil engineers. To realistically asses the structural vulnerability and to incorporate the effects of uncertainties involved in the load-structure system, a probabilistic framework should be utilized. The main components of this framework are presented in Fig. 1.

From a historical point of view, the horizontal component amplitude of ground motions normally plays a dominant role compared to the vertical counterpart. However, acceleration records from the (1989) Loma Prieta earthquake and the (1994) Northridge earthquake in the USA, the (1995) Hyogoken-Kobe earthquake in Japan, (2003) Bam earthquake in Iran, and the (2011) Christchurch earthquake in New Zealand, among others showed that the magnitudes of the vertical component can be as large as, or exceed, the horizontal component. The report from

Elnashai et al. (1995) also highlighted cases of brittle failure induced by direct compression, or by reduction in shear strength and ductility due to variation in axial forces arising from the vertical motion in the (1994) Northridge earthquake. In such situations, most existing code specifications assume that the ratio of vertical component of the ground motion to that of the horizontal component (V/H) varies from 1/2 to 2/3, which must be considered unconservative and needs to be investigated.

Recent studies (Bozorgnia and Campbell 2004; Elgamal and He 2004) on horizontal and vertical ground motions have indicated that such a simple approach is not valid and appropriate for the near-fault regions anymore. The main reasons can be categorized as follows:

- The attenuation rate for vertical ground motion is much higher than that of the horizontal ground motion. This rate increase in the far-field areas. Thus, structures built in the near-fault regions experience higher vertical excitations.
- Vertical ground motion includes more high-frequency content than horizontal ground motion. The difference increases with the decrease in the soil stiffness.

It should also be noted that the higher values of V/H ratio do not necessarily imply more energy content on the desired structure. The reason is that the two components may not coincide in time to cause strong interaction effects.

Besides these, many of the current seismic design codes and damage estimation tools do not include the effect of vertical ground motions on the seismic response of structures and especially columns. However, the observed damage on the columns (diagonal shear cracks) during historical seismic events such as the 1994 Northridge earthquake and the 1995 Kobe earthquake was partly attributed to the effect of vertical motions (Broderick et al. 1994; Elnashai et al. 1995). Field and analytical evidence by Papazoglou and Elnashai (1996) indicated that strong vertical earthquakes can cause a significant fluctuation in the axial force in columns, resulting in a reduction in their shear capacity and compression failure of some of the columns. During the 1995 Kobe earthquake in Japan, the RC structures exhibited very high amplifications of the vertical component of more than two times. The main reasons were the low damping mechanism in the vertical direction and the absence of supplement seismic energy dissipating systems in this direction. On the other hand, because of the high stiffness in the vertical direction, a quasi-resonant response was observed in these structures. High-frequency pulses from vertical motion were recognized as the other reason for such a phenomena (AIJ 1995).

Iyengar and Shinozuka (1972) investigated the effect of self-weight and vertical accelerations on the behavior of tall structures. The structures have been idealized as

Fig. 1 Probabilistic framework for the seismic structural assessment considering the effects of vertical ground motion

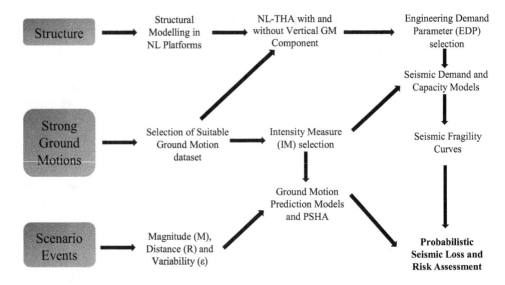

cantilevers and the ground motion as a random process. The main conclusion from their study was that inclusion of self-weight simultaneously with the vertical ground acceleration can increase or decrease the global peak response. These fluctuations in structural response had been considerable in most cases. On the local level, beams were identified as the most critical elements and the effect of vertical ground motion on them had been pronounced. Iyengar and Sahia (1977) investigated the effect of vertical ground motion on the response of cantilever structures using the mode superposition method; their main conclusion is that the consideration of the vertical component is essential in analyzing towers. Anderson and Bertero (1977) used numerical methods to evaluate the inelastic response of a ten-story unbraced steel frame subjected to a horizontal component of earthquake and to combinations of this component with the vertical one; they deduced the following points; the inclusion of the vertical motion on one hand does not increase the displacements, but on the other increases the girder ductility requirement by 50 % and induces plastic deformations in columns. Mostaghel (1974) and Ahmadi (1978, 1980) studied the effect of vertical motion on columns and tall buildings which have been idealized as cantilevers, using the mathematical theory of stability of Liapunov. Their main conclusion was that, in the inelastic region, if the maximum applied earthquake loading would be less than the Euler buckling load, it is guaranteed that the column will remain stable irrespective to the type of earthquake loading, and the inclusion of vertical ground excitation can be neglected. But this is unlikely to be the case for reinforced concrete columns because of the crushing of concrete in compression and the buckling of the yielded reinforcement.

Munshi and Ghosh (1998) investigated the seismic performance of a 12-story RC building under a combination of horizontal and vertical ground motions. This analysis showed a slight increase in the maximum deformation when the vertical ground motion was included. The formation patterns revealed that vertical accelerations induced a slightly different hinge formation pattern and hinge rotation magnitude, and the response of the frame–wall system did not show sensitivity to the vertical acceleration in this case. Antoniou (1997) studied the effect of vertical accelerations on RC buildings by analyzing a eight-story reinforced concrete building designed for high ductility class in Euro code (EC8) with a design acceleration of 0.3 g. This analysis showed that the vertical ground motion can increase the compressive forces by 100 % or even more and lead to the development of tensile forces in columns. These fluctuations in axial forces can result in shear failure in these elements.

Kim et al. (2011) studied the effect of various peak *V/H* ground acceleration ratios and the time lag between the arrival of the peak horizontal and vertical accelerations on the inelastic vibration period and column response for infrastructures. It was observed that the inclusion of vertical motions notably influenced the inelastic response vibration periods and considerably increased or decreased the lateral displacement. It was also noticed that the arrival time had a minimal effect on the axial force variation and shear demand.

None of the previous studies have investigated the code-conforming RC-MRFs utilizing fragility curves and reliability methods. As the seismic vulnerability assessment of high-rise structures is a complex task, it is important to consider that both the lower and higher structural modes might be excited, because of the wide range of frequency content of the applied earthquake loads. On the other hand, the imposed displacements to these structures can be very significant, since the fundamental period of many high-rise structures are within the period range of 1–5 s, which corresponds to the peak displacement spectra of the standard earthquakes. To this end, in the current study, both distributed fiber-based and lumped plasticity approaches and various modes of collapse are considered in the simulation process. To show the significance of vertical ground excitations and to get the most accurate results, a new definition for *V/H* ratio and an optimum intensity measure are proposed. Various mass distributions are considered in the eigenvalue analysis to determine the most accurate and computationally efficient structural model.

Seismic fragility curves as the main component of the current study can be derived using various approaches: observational, experimental, analytical and hybrid techniques to quantify damage and estimate monetary losses (Calvi et al. 2006). While the observational method is the most realistic and rational one, as the entire inventory is taken into consideration it is usually difficult to be utilized because no or insufficient observational-based date are available from the past events. The experimental method is not a feasible option in many cases, because of its cost and the time needed, since a wide range of structures should be tested. In the current study, the third approach based on extensive nonlinear analytical simulations is adopted. This option is the most feasible and possible methodology which can be used in many cases.

The main objective of this study is to calculate the collapse margins and mean annual frequency of collapse as the performance metrics employing displacement-based fragility curves for multiple limit states from concrete cracking to structural collapse in the near-fault areas. The collapse of structures is determined on the basis of the global failure mechanism of the structural system rather than the failure of a structural element. To achieve this goal, numerical models that capture the axial–shear–flexural behavior of the columns are created in nonlinear seismic simulation platforms, Zeus-NL and OpenSees (Elnashai et al. 2004; McKenna 2014).

Selection and characterization of input ground motions

A major stage in the process of fragility estimation is to use appropriate ground motion (GM) records. If the ground motion selection would be done in a way that the hazard consistency is ensured, then the results from the corresponding simulation and analysis would be rational. According to the recommendations of FEMA P-695 (2009), 40 earthquake ground motion records at varying hazard levels from 20 earthquake events are selected from the PEER NGA-WEST2 (Ancheta et al. 2012, 2013; Pacific Earthquake Engineering Research Center 2015)

database (Table 1). This database is the most recent and very suitable for adequate fragility analysis.

The criteria for selection of the analyses records include medium to high vertical component, having large magnitude ($M_w \geq 6.0$) and recorded at near-fault rupture distances ($R \leq 25$ km), with a frequency range to excite the periods of vibration of the structure in both horizontal and vertical directions. The response spectra of the selected records, the median of the acceleration response spectra and the dominant periods (T_1) of the reference structures which will be defined afterward are shown in Fig. 2.

Based on the results illustrated in Fig. 2, the vertical component of a ground motion tends to concentrate all its energy content in a narrow, high-frequency band, while the frequency range for the horizontal component is much wider. Hence, this phenomenon will amplify the structural responses in the short period range, which usually coincide with the vertical periods of RC elements/structures. After the GMs selection, they are amplitude scaled, using the procedure outlined in ASCE 7 (ASCE 2010) to match the 5 % damped site-specific target spectrum (corresponding to the maximum credible earthquake, MCE) given in Fig. 3.

Representative set of structures

Four RC-MRFs ranging from 7 to 20 stories are selected to represent medium- and high-rise buildings. The frames are designed and detailed according to ACI building code ACI-318 (2011) and ASCE 7 (2010) provisions. Two categories of RC-MRFs, special and intermediate, are used in the current study. The ordinary MRFs, because of their low level of ductility during an earthquake, are not considered here. The special MRF employs the strong column weak beam (SCWB) concept and specifies elaborate detailing of joints. Thus, the SMRF is expected to form the sway mechanism and possesses a high degree of ductility. On the other hand, the intermediate MRF has enough strength as well as reasonable ductility and can be used throughout most of the seismic-prone areas. 7- and 12-story buildings are designed as intermediate MRFs, while 15- and 20-story buildings are designed as special MRFs. The behavior factors (*R*) are considered as 5 and 8, respectively (ASCE 7 2010).

Span lengths are identical in both directions equal to 6 m, and story heights are 3.5 m. Damping is set to be 5 % in the first three modes and is considered as of Rayleigh mass and stiffness proportional type, based on the recommendations of Zareian and Medina (2010). The reference structures are shown in Fig. 4 and the beam and column dimensions and longitudinal and transverse steel ratios are listed in Table 2.

Table 1 Input ground motions used for the nonlinear response history analyses (NL-RHA)

No.	Earthquake name	Date	Station name	Moment magnitude, M_w	Epicentral distance, (km)	$(PGA)_H$, g	$(PGA)_V$, g
1	Wenchuan, China	2008	Wenchuanwolong	7.90	19.54	0.77	0.96
2	Chi-Chi, Taiwan	1999	TCU078	7.62	4.96	0.38	0.17
3	Chi-Chi, Taiwan	1999	TCU089	7.62	14.16	0.75	0.34
4	Chi-Chi, Taiwan	1999	TCU079	7.62	15.42	0.59	0.42
5	Kocaeli, Turkey	1999	Izmit	7.51	5.31	0.19	0.14
6	Kocaeli, Turkey	1999	Yarimca	7.51	19.30	0.29	0.24
7	Tabas, Iran	1978	Dayhook	7.35	20.63	0.33	0.19
8	Landers, USA	1992	Joshua Tree	7.28	13.67	0.27	0.18
9	Landers, USA	1992	Morongo Valley Fire Station	7.28	21.34	0.19	0.16
10	Duzce, Turkey	1999	Duzce	7.14	1.61	0.43	0.35
11	Duzce, Turkey	1999	Lamont 1058	7.14	24.05	0.68	0.19
12	Duzce, Turkey	1999	IRIGM 487	7.14	24.31	1.00	0.33
13	Golbaft, Iran	1981	Golbaft	7.00	13.00	0.28	0.24
14	Darfield, New Zealand	2010	GDLC	7.00	4.42	0.73	1.25
15	Darfield, New Zealand	2010	HORC	7.00	10.91	0.47	0.81
16	Loma Prieta, USA	1989	Corralitos	6.93	7.17	0.50	0.46
17	Loma Prieta, USA	1989	BRAN	6.93	18.46	0.59	0.90
18	Loma Prieta, USA	1989	Capitola	6.93	20.35	0.44	0.14
19	Kobe, Japan	1995	Nishi-Akashi	6.90	8.70	0.47	0.39
20	Kobe, Japan	1995	IWTH26	6.90	13.12	0.67	0.28
21	Kobe, Japan	1995	Takatori	6.90	19.25	0.32	0.57
22	Nahanni, Canada	1985	Site 2	6.76	6.52	0.40	0.67
23	Nahanni, Canada	1985	Site 1	6.76	6.80	1.16	2.28
24	Northridge, USA	1994	Rinaldi Receiving	6.69	5.41	1.64	1.05
25	Northridge, USA	1994	Arleta-Nordhoff Fire Sta	6.69	8.48	0.75	0.32
26	Northridge, USA	1994	LA Dam	6.69	20.36	1.39	1.23
27	Niigata, Japan	2004	NIG019	6.63	4.36	1.26	0.80
28	Niigata, Japan	2004	NIG020	6.63	21.52	1.48	0.57
29	Bam, Iran	2003	Bam	6.60	12.59	0.74	0.97
30	Zarand, Iran	2005	Zarand	6.40	16.00	0.31	0.30
31	Imp. Valley, USA	1979	Bonds Corner	6.53	6.19	0.69	0.53
32	Imp. Valley, USA	1979	Calexico Fire Station	6.53	19.44	0.31	0.25
33	Imp. Valley, US	1979	Chihuahua	6.53	24.82	0.17	0.21
34	Christchurch, New Zealand	2011	Heathcote Valley Primary School	6.20	1.11	1.39	2.18
35	Christchurch, New Zealand	2011	LPCC	6.20	4.89	0.65	1.90
36	Morgan Hill, USA	1984	Halls Valley	6.19	16.67	0.35	0.21
37	Morgan Hill, USA	1984	Zack Brothers Ranch	6.19	24.55	0.94	0.39
38	Talesh, Iran	1978	Talesh	6.00	15.00	0.23	0.13
39	Parkfield, USA	2004	Parkfield-Stone Corral 1E	6.00	7.17	0.72	0.33
40	Parkfield, USA	2004	Parkfield-Stone Corral 2E	6.00	9.28	0.83	0.72

Structural modeling approaches

In general, most of the current models provided by other researchers cannot be utilized to predict the accurate behavior of the RC elements in the presence of vertical excitations. The main reason is that such models do not account for important response features, such as the interaction between shear, flexure and axial forces. One of the main failure modes of RC columns due to vertical excitations as mentioned previously is the shear failure in

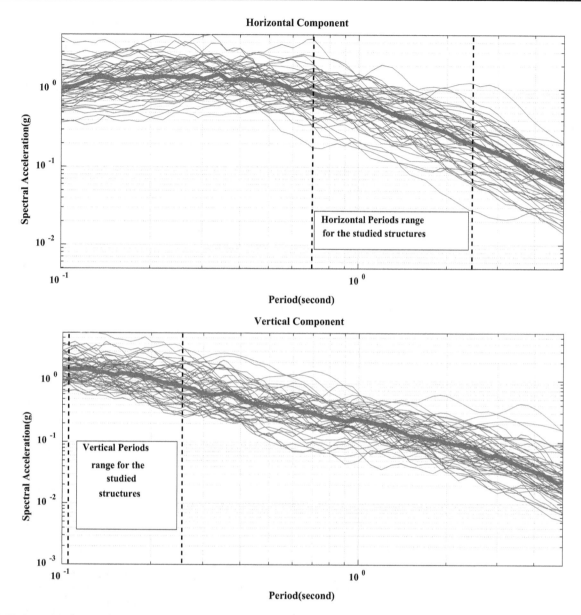

Fig. 2 Horizontal and vertical response spectra of the selected ground motions

these members. To include this type of failure in the analytical models, some modification should be incorporated in the modeling approaches. To this end, two nonlinear simulation platforms; ZEUS-NL (Elnashai et al. 2004) and OpenSees (McKenna 2014), are used to simulate the seismic response of the reference structures. Fixed-base models are used in the analysis stage; as a result, soil–structure foundation interaction is neglected. A leaning column to account for the P-Δ effect from loads on the gravity system is also considered in both platforms.

In ZEUS-NL, the columns are modeled with distributed nonlinear fiber sections (flexural response) and an NL zero-length shear spring (shear response) at the end of each column. The idealization adopted in the first approach

effectively models reinforcing steel, and unconfined and confined concrete (Mander et al. 1988). This approach allows monitoring the stress–strain response at each fiber over several Gauss sections through the integration of the nonlinear stress–strain response of different fibers in which the section is subdivided, as shown in Fig. 5 (shorter fiber-based elements at the end of the column and longer fiber-based elements away from the end of the column are used to capture inelastic flexural response in plastic hinge zones). This modeling approach reduces the modeling uncertainty. Material properties used from the large database of ZEUS-NL for including the damage plasticity in both concrete and reinforcements. "*stl1*" material model is chosen for reinforcement steel, which is a bilinear elasto-

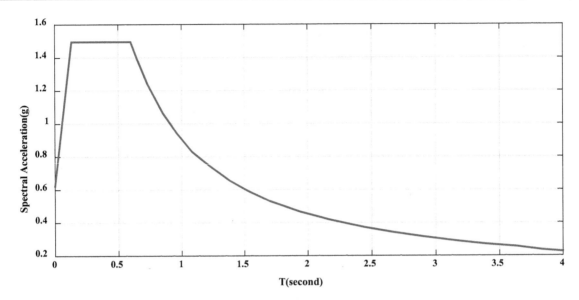

Fig. 3 MCE level spectra for the maximum seismic design category (MSDC)

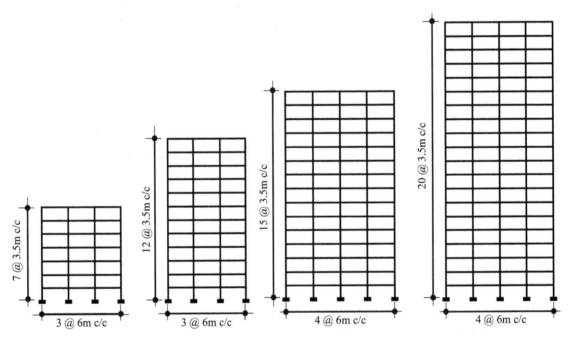

Fig. 4 Schematic presentation of the case studies

plastic ideal model with kinematic strain hardening; "*con2*" material model is chosen for the concrete, which is a uniaxial concrete model that includes the confinement provided by hoops or ties.

To simulate the shear response of the columns Lee and Elnashai (2001, 2002) developed two shear models: hysteretic shear model under constant axial force (Lee and Elnashai 2001) and hysteretic shear model under axial force variation (Lee and Elnashai 2002). In case of vertical excitations, as the axial force is not constant, the first shear model cannot capture the effect of axial force

variation on the seismic response. Figure 6a shows the envelope curve of the hysteretic shear model under constant axial force, which is defined by a quadrilinear symmetric relationship comprising cracking, yielding and ultimate conditions. The response parameters on the curve are determined by the modified compression field theory (MCFT) developed by Vecchio and Collins (1986). To address the effect of varying axial force on the envelope curve, the basic formulation can be extended and the curves extracted for multiple levels of axial force, as shown in Fig. 6b.

Table 2 Frame element sizes and reinforcements details

Member specifications	Reference structures and stories range											
	20S4B				15S4B			12S3B			7S3B	
	1–5	6–10	11–15	16–20	1–5	6–10	11–15	1–4	5–8	9–12	1–4	5–7
Beam												
b (cm)	60	50	45	35	55	45	35	50	40	35	45	35
h (cm)	90	70	60	45	75	60	45	60	50	45	50	45
ρ_l (%)	2.2	2.2	1.8	1.8	2.0	1.9	1.7	2.0	2.0	1.9	2.2	2.0
ρ_t (%)	1.0	1.0	1.0	1.0	1.0	1.0	1.0	1.0	1.0	1.0	1.0	1.0
Exterior column												
b (cm)	100	80	65	50	85	65	50	65	55	45	50	40
h (cm)	100	80	65	50	85	65	50	65	55	45	50	40
ρ_l (%)	2.1	1.9	1.9	1.8	2.0	1.8	1.8	2.1	1.8	1.7	1.9	1.8
ρ_t (%)	2.0[a]	1.4	1.4	1.4	2.0[a]	1.4	1.4	2.0[a]	1.4	1.4	2.0[a]	1.4
	1.4				1.4			1.4			1.4	
Interior column												
b (cm)	120	90	75	60	95	75	55	75	60	50	60	45
h (cm)	120	90	75	60	95	75	55	75	60	50	60	45
ρ_l (%)	1.8	1.7	1.5	1.5	1.8	1.5	1.5	1.7	1.7	1.6	1.9	1.7
ρ_t (%)	2.0[a]	1.6	1.6	1.6	2.0[a]	1.6	1.6	2.0[a]	1.6	1.6	2.0[a]	1.6
	1.6				1.6			1.6			1.6	

[a] First-story columns only

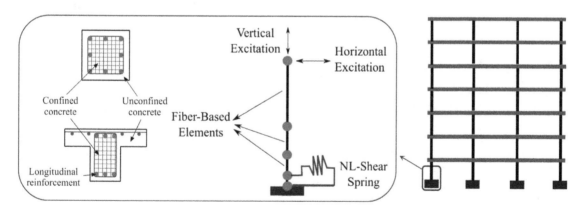

Fig. 5 First modeling approach used in ZEUS-NL

A fully concentrated plasticity (Krawinkler et al. 2004) approach is utilized in OpenSees to consider both the flexural and shear behavior of the columns in terms of vertical excitations. To this end, all the beams, columns and joints are modeled with NL springs. Beams and joints are modeled with rotational springs and the columns are modeled using rotational springs along with a zero-length shear spring located at one end of the column to consider the effect of shear failure as shown in Fig. 7. The plastic hinge behavior of the beam and column elements are calibrated through the large set of experimental data by Haselton and Deierlein (2007) and Haselton et al. (2008).

The hysteretic uniaxial material model monitors the shear/flexural behavior of the columns. In this model, shear deformations are simulated using a shear spring, while the flexural deformations are monitored using beam/column elements. As illustrated in Fig. 8, the solid line indicates the total response of the RC column. This line captures the degrading shear behavior, when the shear strength would be less than the shear corresponding to the plastic hinges development. A dashed line is presented in Fig. 8, which shows the case in which the shear strength is higher than the shear corresponding to the plastic hinges development and the model does not capture any shear degradation.

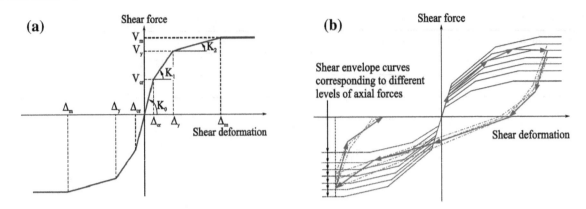

Fig. 6 Schematic presentation of zero-length shear spring envelope curve used in ZEUS-NL

Fig. 7 Second modeling approach used in OpenSees

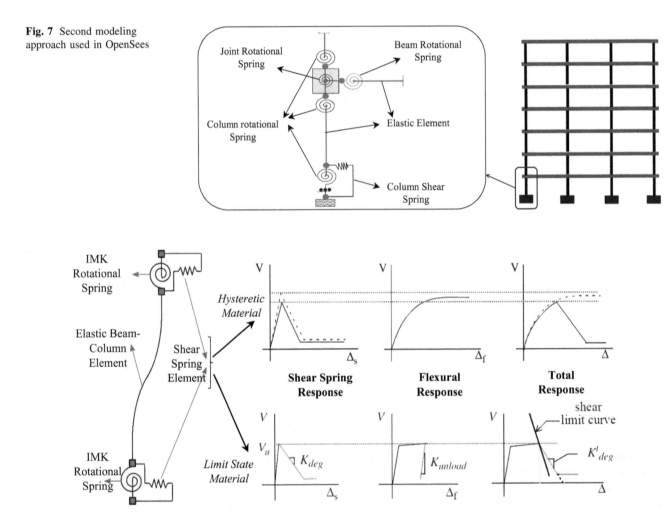

Fig. 8 Schematic representation of shear springs in series model using hysteretic limit-state material model

Watanabe and Ichinose (1992), Aschheim and Moehle (1992), Priestley et al. (1994) and Sezen (2002) have shown that the shear strength in RC elements decays with increased plastic deformations. Hence, the dashed line provided in Fig. 8 cannot be realistic and accurate, if the column yields in flexure close to its estimated shear strength. The main deficiency of the hysteretic uniaxial model is that it determines the point of shear failure based only on the column shear, while it should be determined based on both force and deformation.

To resolve this problem, the shear load versus deformation model for the shear spring was developed using the existing OpenSees limit-state material model and shear limit curve developed by Elwood (2004). As shown in Fig. 8, the shear limit curve is activated and shear failure is initiated once the column shear demand exceeds the column shear capacity. In this case, the limit-state material model simulates and captures the RC column response to detect the possibility of shear failure. To this end, the shear limit curve is defined according to both the column shear and the total displacement or drift ratio (Fig. 8). In case the columns are vulnerable to shear failure after flexural yielding, then the drift capacity model proposed by Elwood and Moehle (2005) can be utilized to define the accurate limit curve. Other shear failure criterion such as plastic rotation at the two ends of the column is also proposed by LeBorgne (2012).

The beams' and columns' moment–rotation behavior is simulated utilizing an NL hinge model including strength and stiffness deterioration developed by Ibarra et al. (2005) and has been used in PEER/ATC (2010) as well. As the reference structures are code conforming and have ductile behavior, the peak-oriented model is utilized. This model combines a post-peak negative stiffness branch of the backbone curve to capture in-cycle strain softening and a cyclic model to capture the strength and stiffness deterioration based on the cumulative hysteretic energy dissipated (Fig. 9). The individual modes of deterioration are described briefly below:

Basic strength degradation This mode of deterioration is defined in a way to show a reduction in the yield strength. This mode also includes the strain hardening slope degradation. Based on the formulation, the values for positive and negative directions are defined independently (Fig. 9a).

Post-capping strength degradation It is defined by translating the post-capping branch toward the origin; however, the branch slope will keep constant and it will move inward to reduce the reference strength. The process will be done each time the X axis is crossed (Fig. 9b).

Unloading stiffness degradation This mode of deterioration indicates the reduction in both positive and negative unloading stiffness. The unloading stiffness ($K_{u,i}$) in each cycle depends on the unloading stiffness in the previous excursion ($K_{u,i-1}$) (Fig. 9c).

Accelerated reloading stiffness degradation This mode of deterioration escalates the target displacement according to the loading trend in both positive and negative directions (Fig. 9d).

For ductile RC frames, beam–column joints are often modeled with rigid joint zones. However, Shin and LaFave (2004) argued that the joint regions are not fully rigid and may experience some shear deformations that can contribute to global deformation. Therefore, the analytical model must predict the inelastic behavior of ductile beam–column joints. In the current study, the pinching material model is utilized for the joint rotational springs. The modeling parameters of

the pinching material for ductile exterior and interior beam–column joints have been calibrated by Jeon (2013) and used in the modeling procedure (Fig. 10).

Eigenvalue analysis

The detailed analytical models were subjected to eigenvalue analyses to determine the fundamental period of the structures (Chopra 1995). Zero or small errors for the horizontal modes imply that distributing the lumped-mass mesh over beams constitutes no differences, but for the case of vertical excitation the results show a lot of discrepancies. The vertical and horizontal periods for several lumped-mass models and a distributed one are calculated and compared (Fig. 11; Table 3). It is then assumed that if a simplified lumped-mass model gives very similar natural mode shapes and vibration periods compared to the distributed one, this simplified model is reliable to simulate vertical motions and can be used in the NL-RHA.

Compared to L-Mass 1, L-Mass 2 is much more accurate. However, there are clearly differences in the vertical periods. L-Mass 3 and L-Mass 4 show both significantly similar periods in the horizontal and vertical directions. Even though L-Mass 3 has a coarse lumped-mass approach compared to the distributed mass model, the eigenvalues of this model imply very similar tendencies to the exact solution. As a result, because of the small differences in L-Mass 3 compared to L-Mass 4 results, L-Mass 3 is the most simplified lumped-mass model to cover realistic vertical motion with minimum computational effort and is implemented in the NL-RHA of the studied structures to provide fragility curves.

V/H ratio (conventional and proposed approaches)

The current and most common design practice is to take the *V/H* spectral ratio as 2/3 as proposed by Newmark et al. (1973) and is also used by FEMA (Bozorgnia and Campbell 2004). However, this approach is inaccurate for near-source moderate and large earthquakes (Friedland et al. 1997; Bozorgnia et al. 1999; Bozorgnia and Campbell 2004; Button et al. 2002). To clarify this issue, a large set of earthquake records were extracted from NGA-WEST2 (Ancheta et al. 2012, 2013; Pacific Earthquake Engineering Research Center 2015) dataset to compare the peak vertical to horizontal ground acceleration (*V/H* ratio) at different ground motion magnitudes and distances. Figure 12 shows the detailed comparison. For a better illustration, a linear trend line is fitted to all data points.

The results indicate that the 2/3 rule is unreasonable and confirms that the *V/H* ratio may show significant variations, which depend on the source and site characteristics and

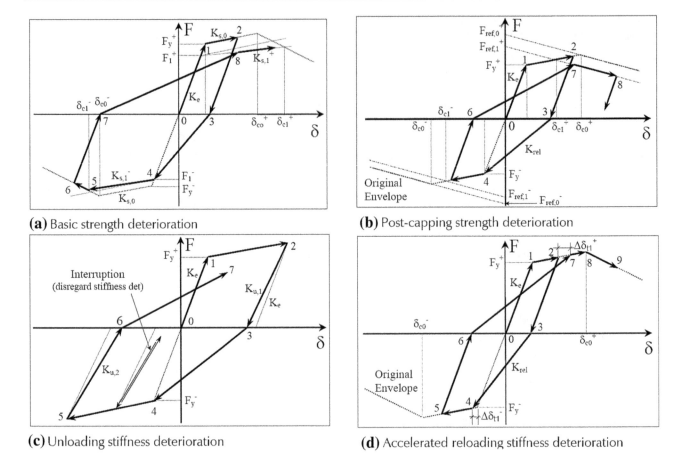

(a) Basic strength deterioration

(b) Post-capping strength deterioration

(c) Unloading stiffness deterioration

(d) Accelerated reloading stiffness deterioration

Fig. 9 Individual deterioration modes illustrated for a peak-oriented model (Ibarra et al. 2005)

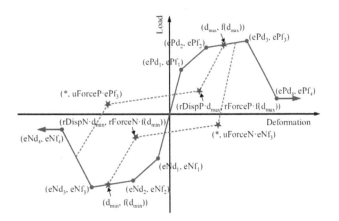

Fig. 10 Constitutive pinching material model proposed by Lowes and Altoontash (2003)

seismic radiation pattern. The shortage of the conventional *V/H* calculation methods is that they do not differentiate the horizontal and vertical frequency content and cannot include the influence of dynamic properties of the considered structure. To this end, a new approach is proposed by the authors, which calculates the *V/H* based on the dominant vertical and horizontal spectral acceleration of the

studied structures. NGA-WEST2 (Ancheta et al. 2012, 2013; Pacific Earthquake Engineering Research Center 2015) dataset composed of more than 20,000 records is used to provide reliable calculations. Based on the eigenvalue analyses, the horizontal and vertical dominant periods of the case studies were evaluated and used to calculate the vertical to horizontal ratio for each structure exclusively. The benefit of the new approach compared to the conventional methods is that for each structure with a specific structural system, the *V/H* will be calculated based on its dynamic properties and would be unique. The results for the new approach are presented in Fig. 13. Red lines indicate the central 50 % (median), while lower and upper boundary lines show the 10 and 90 % quantile of the data. Two vertical lines extend from the central box, indicating that the data remaining outside the central box extends maximally to 1.5 times the height of the central box, but not past the range of the data. The reaming data points plotted by red markers are the outliers. The given results show that there is a correlation between the structure's height and *V/H* ratio, and the high-rise RC-MRFs experience higher levels of seismic vertical excitations. However, it does not prove that these structures may be more

Fig. 11 Various lumped-mass models considered in the eigenvalue analysis

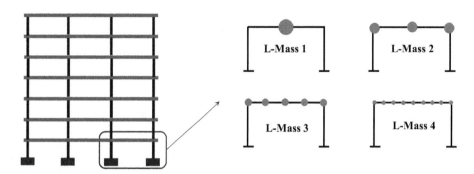

Table 3 Vertical and horizontal periods of the case studies considering various mass models

Reference structures	Mass model	Horizontal period (s), T_h	Vertical period (s), T_v
7 Story	L-Mass 1	0.79	0.36
	L-Mass 2	0.79	0.17
	L-Mass 3	0.77	0.13
	L-Mass 4	0.76	0.12
	Distributed mass	0.74	0.11
12 Story	L-Mass 1	1.38	0.41
	L-Mass 2	1.35	0.22
	L-Mass 3	1.35	0.18
	L-Mass 4	1.34	0.17
	Distributed mass	1.32	0.15
15 Story	L-Mass 1	1.81	0.53
	L-Mass 2	1.81	0.29
	L-Mass 3	1.79	0.22
	L-Mass 4	1.78	0.19
	Distributed mass	1.78	0.18
20 Story	L-Mass 1	2.53	0.67
	L-Mass 2	2.51	0.38
	L-Mass 3	2.50	0.29
	L-Mass 4	2.49	0.27
	Distributed mass	2.45	0.24

vulnerable when subjected to multi-component excitations. A detailed investigation is given in the following sections.

Performance criteria

In this study, interstory drift ratio (IDR) is considered as the engineering demand parameter (EDP). This is particularly a suitable choice for RC-MRFs, since it relates the global response of the structure to joint rotations where most of the inelastic behavior in the moment-resisting frames is concentrated.

At the first stage, limit states are considered corresponding to different performance levels as specified in FEMA 356 (2000). The IDR values of 1, 2 and 4 % are used for IO, LS and CP performance levels, respectively. In the next stage, these performance levels are calibrated and verified for each reference structure, based on its structural capacity through nonlinear pushover analysis. The IDR limit for each individual structure from pushover analysis combining of the first four natural modal shapes, weighted by modal mass participation factors, is used for the lateral load distribution pattern. IDR values corresponding to the first concrete cracking/steel yielding, concrete strain corresponding to maximum confined concrete stress and maximum confined concrete strain are considered as IO, LS and CP performance levels, respectively. Table 4 lists the median capacity values against each performance level for all buildings. The results show good agreement of the NL-pushover capacities with FEMA-356 performance levels.

Besides the limit states mentioned above, the collapse limit state can be defined on the basis of a different approach. The onset of 'collapse' for a ground motion record is identified as the point where maximum IDR response increases 'drastically' when the spectral acceleration of the record is increased by a 'small' amount (Vamvatsikos and Cornell 2002). To this end, in the next stages, incremental dynamic analysis (IDA) is performed on the reference MRFs and the collapse is defined as the point of dynamic instability when IDR increases without bounds for a small increase in the ground motion intensity for each structure individually.

Proposing an optimum intensity measure for fragility analysis

The ground motions are characterized by intensity measures (IMs), and their choice plays a crucial role in the seismic fragility estimation. An optimal IM is the one that has good efficiency, sufficiency, practicality, hazard computability and predictability among other characteristics (Mackie and Stojadinovic 2001; Luco and Cornell 2007; Giovenale et al. 2004; Padgett et al. 2008). Efficiency means the ability to accurately predict the response of a

Fig. 12 V/H ratio vs. the
epicentral distance (different
magnitude windows)

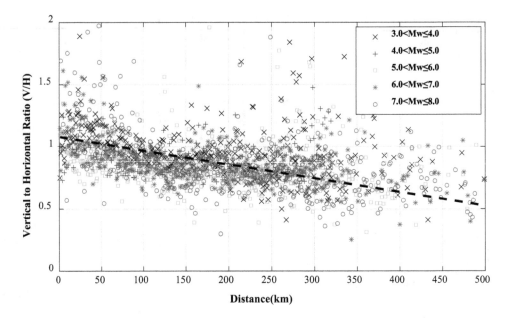

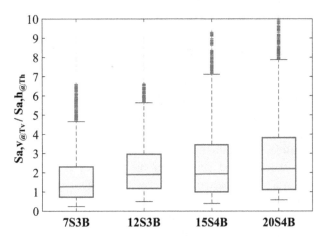

Fig. 13 Box plot of *V/H* ratios based on dominant spectral acceleration of the studied structures

structure subjected to earthquakes (i.e., small dispersion of
structural response subjected to earthquake ground motions
for a given IM). A sufficient IM is defined as one that
renders structural responses subjected to earthquake ground
motions for a given IM conditionally, independent of other
ground motion properties (i.e., no other ground motion
information is needed to characterize the structural
response). Previous studies have shown that PGA is not an

accurate and ideal IM for evaluating the geotechnical
phenomenon, as it cannot consider the ground motion
duration (Kramer et al. 2008). It is found that the fragility
curves based on vector-valued IM are better able to represent the damage potential of earthquake (Baker 2015).
Thus, an optimized vector-valued intensity measure, which
includes the geometric mean of spectral accelerations over
a range of period, is considered in the current study. The
parameters in the vector-valued intensity measure should
be chosen to convey the most possible information between
the ground motion hazard and the structural response
stages of analysis. This requires identifying parameters that
most affect the structure under consideration.

The adopted IM is a suitable choice, especially for the
high-rise RC-MRFs where the effect of higher modes is
significant. Given that $S_a(T_1)$ has been verified as an
effective predictor of structural response for a wide class of
structures, it will be used as the first element of the vector,
while the effect of higher modes (HM) is considered as the
second parameter (Eqs. 1 and 2).

$$\text{IM(vector)} = S_a(T_1) + \text{HM}, \tag{1}$$

$$\text{HM} = \frac{\sqrt[10]{[S_a(k_1T_1) \times \ldots \times S_a(k_uT_1)]}}{S_a(T_1)}, \tag{2}$$

Table 4 Performance levels
according to FEMA-356 and
NL-pushover analysis in terms
of IDR (%)

Reference structures	Immediate occupancy (IO)		Life safety (LS)		Collapse prevention (CP)	
	FEMA-356	Pushover	FEMA-356	Pushover	FEMA-356	Pushover
7S3B	1	0.87	2	2.03	4	3.42
12S3B	1	0.93	2	2.27	4	3.78
15S4B	1	1.12	2	2.29	4	4.08
20S4B	1	1.27	2	2.38	4	4.25

where the constant k_1 is chosen to vary between T_{low}/T_1 and 1, and k_u between 1 and T_{upp}/T_1, where T_{low} and T_{upp} are, respectively, the lower and the upper period of the elastic spectrum. It is worth mentioning that k_1 and k_u values can be optimized for each structure separately based on its dynamic characteristics using a trial and error procedure.

An efficient IM reduces the amount of variation in the estimated demand for a given IM value (Giovenale et al. 2004; Padgett et al. 2008). Employing an efficient IM yields less dispersion about the estimated median in the results of the NL-RHA. The comparison of results in terms of standard deviation for conventional scalar and vector IMs are illustrated in Fig. 14. In the comparison, $S_a(T_1)$ is considered as the most conventional scalar IM, while $[S_a(T_1), S_a(T_2)/S_a(T_1)]$ is considered as a conventional vector IM.

As can be seen from Fig. 14, $S_a(T_1)$ can be an efficient predictor for structures with short period, but not for the structures with medium to large periods. On the other hand, the vector IM can predict the results in an efficient way for both short and large period structures; however, the results have more standard deviations compared to the proposed IM. In all cases, IM (New) has been more efficient and sufficient, as it is always associated with small values of dispersion. The maximum dispersion values are 20, 31 and 50 % for IM (New), IM (Vector) and $S_a(T_1)$, respectively.

Probabilistic demand models

To formulate and correlate the earthquake intensity measure (IM) to the building-specific demand measure (DM), a probabilistic seismic demand model (PSDM) should be developed. Selecting an appropriate pair of IM and DM is a difficult task, as the PSDM results are practically sufficient and efficient (Gardoni et al. 2002, 2003). In the current study, response history analyses (RHAs) are used to

quantify the seismic demand of the reference structures. The maximum interstory drift ratio (Max IDR) is considered as a suitable DM, since it is closely related to damage states of the RC-MRFs. The final results are illustrated in terms of IDR-IM in Fig. 15, under 40 sets of ground motions chosen in the previous sections.

NL-incremental dynamic analyses and seismic fragility estimate

In the fragility estimation process, a suitable analysis procedure should be implemented. Based on the recommendations provided by Vamvatsikos and Cornell (2002) and FEMA P-695 (2009), incremental dynamic analysis (IDA) is recognized as one of the best and most common procedure, where a suite of earthquake records is scaled repeatedly to find the intensity measure in which the structural collapse will occur. In this study to isolate the effect of each component, horizontal and vertical ground motions were applied separately and simultaneously.

Using the IDA approach, information about variability in ground motions can be directly incorporated into the collapse performance assessment. However, this process only captures the record-to-record variability and does not account for how well the nonlinear simulation model represents the collapse performance of the reference structures; hence, model uncertainties should also be accounted in the collapse simulation, which will be discussed later.

Effect of vertical excitations on the structural responses (local level)

Seismic performance evaluation of RC structures subjected to (medium to strong) multi-component ground motions has some complexities. One of these difficulties is the increment in axial force variation in RC columns, which

Fig. 14 Comparison of the results' dispersion for different IM

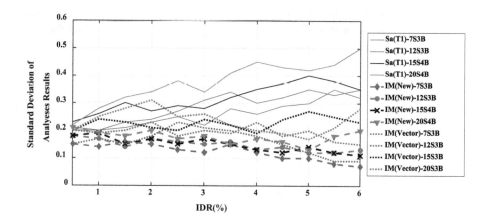

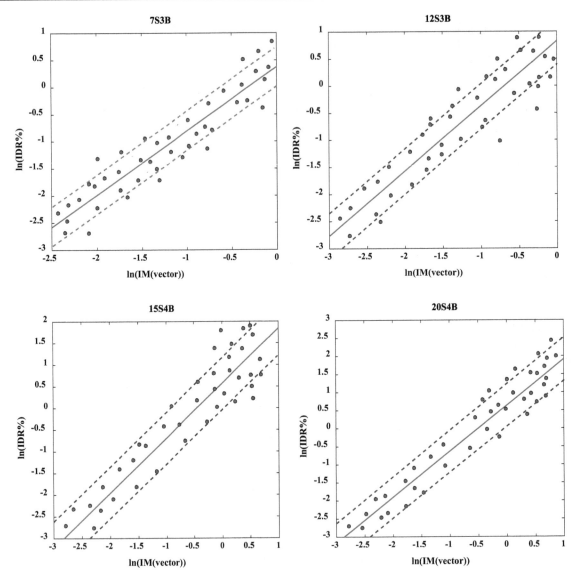

Fig. 15 Probabilistic seismic demand models for the reference structures

can be superimposed in the overturning forces. Since there is a direct relation between flexural, shear and axial forces, the fluctuation in axial force increases the possibility of shear failure. This is mainly due to the significant variation in strength and stiffness in the columns caused by vertical excitation. NL-RHA was performed using the selected natural horizontal ground motions applied with and without the vertical component.

Seismic loading may be applied upward as well as downward, thus subjecting structural members to unaccounted-for action, if the naturally existing vertical ground excitations are neglected in the design procedures. Among the structural members, columns are the most vulnerable and may be adversely affected, if high compressive forces are developed or if axial forces change to tension. Sample results of the effect of vertical ground motions on the first-story column of the 7S3B reference model are shown in Fig. 16.

Table 5 provides the ratio of the vertical seismic force (maximum axial force in the column) to axial gravity load on the column with and without vertical ground motion for the reference structures for three V/H ratios in the first-story interior columns. From the obtained results, it is observed that the effect of vertical ground motion on the ratio of axial force to gravity load increases significantly for all the models in the range of 4.19 and 108.59 % with increase of the V/H ratio.

Another complexity which occurs in the presence of high vertical forces is the increment in ductility demand and reduction in ductile capacity; which may result in extensive damages. The shear capacity may be significantly affected by the variation in column axial forces, which may cause loss or reduction of the axial load contribution to shear strength. As shown in Fig. 17, the shear capacity of an interior first-story column in 7S3B

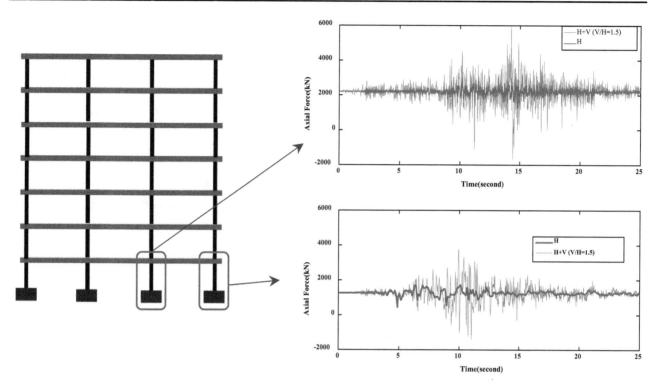

Fig. 16 Column axial force variation in the presence of vertical ground motion (*V/H* ratio = 1.5)

Table 5 Averaged ratio of vertical seismic force to gravity load force for all the structures

Reference structure	*V/H* ratio	Horizontal only	Combined (*H* + *V*)	Increment due to vertical excitations (%)
7S3B	1.0	1.53	1.92	25.49
	1.5	1.62	2.45	51.23
	2.0	1.28	2.67	108.59
12S3B	1.0	1.92	2.18	13.54
	1.5	1.68	1.92	14.29
	2.0	2.15	2.24	4.19
15S4B	1.0	1.78	2.10	17.98
	1.5	1.93	2.19	13.47
	2.0	1.67	2.78	66.47
20S4B	1.0	1.78	2.27	27.53
	1.5	1.95	2.34	20.00
	2.0	2.52	2.95	17.06

frame was calculated using ACI-318 (2011) and ASCE-41 (2013) formulas and compared to shear demand with and without vertical ground motion. The priority of ASCE-41 (2013) formula over the ACI-318 (2011) formula is that it is based on the displacement ductility demand, which is more realistic. It is very evident that in the case of VGM, the shear demand exceeds the shear capacity. Hence, the

possibility of shear failure would be increased before the structure reaches the global collapse state. It is worth mentioning that similar results are observed for all the studied MRFs.

Seismic fragility estimation (global level)

For the case of earthquake excitations, a closed-form solution is not usually available, since there are a large number of random variables and different probability density functions associated with these events. To resolve this problem, the reliability of the structures under these complex phenomena can be represented using a probabilistic methodology incorporating fragility curves. These curves define the probability of exceeding a specific damage state subjected to a hazard by a suitable IM.

In this study, following the structural reliability theory by Ditlevsen and Madsen (1996), the fragility functions are formulated as in Eq. (3):

$$F(\text{IM}; \eta) = P[\{g(C, \text{IM}; \eta) \leq 0\}|\text{IM}].\tag{3}$$

Based on the inherent randomness and uncertainty in the capacity (*C*), the seismic demand (*D*) and the limit states, a closed-form approximation using a log-normal distribution was proposed by Wen et al. (2004). This fragility formulation is given as:

Fig. 17 Shear supply and
demand for 7S3B building for
H and *H* + *V* under a record
with *V/H* = 1.5

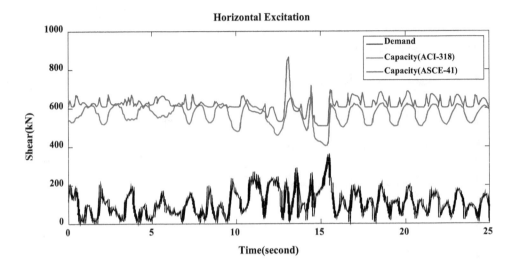

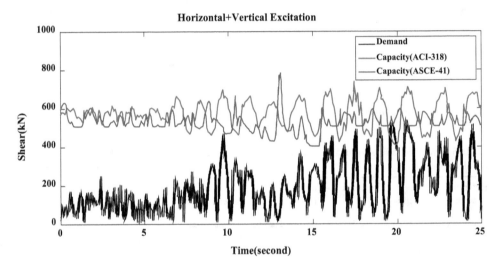

$$F(\text{IM}; \eta) \cong 1 - \Phi\left(\frac{\chi_\text{C} - \chi_\text{D|IM}}{\sqrt{\beta_\text{C}^2 + \beta_\text{D|IM}^2 + \beta_\text{M}^2}}\right), \qquad (4)$$

where $\Phi(.)$ denotes the standard normal CDF, and χ_C and $\chi_\text{D|IM}$ are the natural logarithm of the median capacity and demand of the system, respectively, while β_C, $\beta_\text{D|IM}$ *and* β_M represent the uncertainty in estimating the capacity, demand and structural modeling. Fragility estimates for the studied MRFs are obtained using the IDA results for the optimized HM values and are presented in Figs. 18 and 19. Both modeling approaches are included in the fragility estimation and the results show that two approaches match very well and less than 10 % dissimilarity was observed in all performance levels. For a better demonstration of the collapse mode under concurrent horizontal–vertical excitations, the collapse fragility surfaces are plotted in a 3D space as well (Fig. 20).

Based on Figs. 18 and 19, when the vertical component is coupled with the horizontal excitations, the fragility results can be changed extensively. The maximum increase in the structure's fragility appeared in the CP and Collapse damage modes. Figure 20 shows the developed fragility surface based on a vector-valued IM and can be visualized as fragility curves by projecting the surface onto the planes. These figures demonstrate the wide variation between fragility curves based on scalar-valued intensity measure.

It can be seen in Fig. 20 that there is a discrepancy of up to 30 % between the curves calculated for various HM values. The main advantage of these fragility surfaces is that the variability of structural fragility due to a second parameter can be accounted for in contrast to when fragility curves are used. This means that which records should be used depends on the seismic hazard at the site when scalar-valued IM [$S_\text{a}(T_1)$] is used to evaluate the structural fragility.

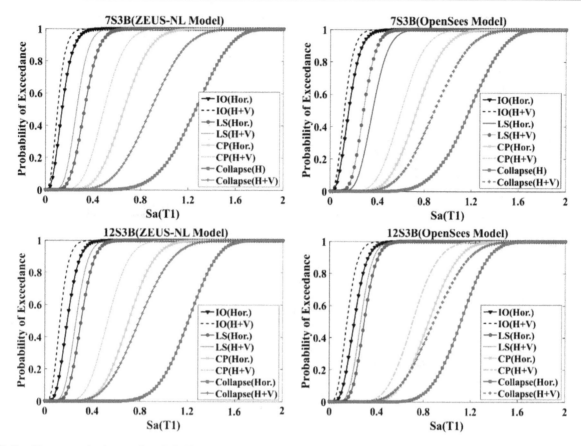

Fig. 18 Fragility curves for intermediate RC-MRFs at the optimal HM

Ignoring the effect of HM will bias the final results. For example, if the seismic hazard disaggregation suggests that extreme motions are associated with records having a mean value of HP of about 0.75, but records are selected with a mean value of about 0.50, then the $S_a(T_1)$-based result will underestimate the seismic fragility. In other words, the evaluation of structural fragility by means of vector-valued IMs reduces the complexity of record selection procedure based on the seismic environment (i.e., magnitude, distance, site conditions, etc.).

Collapse performance evaluation

Prior to the development of incremental dynamic analysis (IDA) and its use in the FEMA P-695 methodology (2009), accurate modeling of buildings near collapse, in the negative post-peak response range, was not a high priority of research. In recent FEMA guidelines (e.g., FEMA P-440A 2009; FEMA P-695 2009), sideway collapse (where collapse is defined based on unrestrained lateral deformations with an increase in ground motion intensity) is typically assumed to be the governing collapse mechanism.

In the current study, important metrics for quantifying collapse resistance of structures are defined in the previous section and illustrated in Fig. 21, including the median collapse capacity and the conditional probability of collapse at an intensity level of interest, the code-defined maximum considered earthquake (MCE). As the differences in two modeling approaches were negligible, results from the ZEUS-NL models which had higher accuracies are utilized in this section.

The MCE intensities are obtained from the response spectrum of MCE ground motions (Fig. 3) at the fundamental period (T_1). The ratio between the median collapse intensity and the MCE intensity is defined as the collapse margin ratio (CMR), which is the primary parameter used to characterize the collapse safety of the structure (Eq. 5):

$$CMR = \frac{S_a(T_1)_{@P[Collapse]=50\%}}{S_a(T_1)_{@MCE}}. \tag{5}$$

Comparing the intermediate and special MRFs located in the same hazard region, Fig. 21 indicates that the 20- and 15-story buildings have better collapse behavior compared to the 7- and 12-story buildings. The reason for this behavior is due to the higher ductility level in SMRFs.

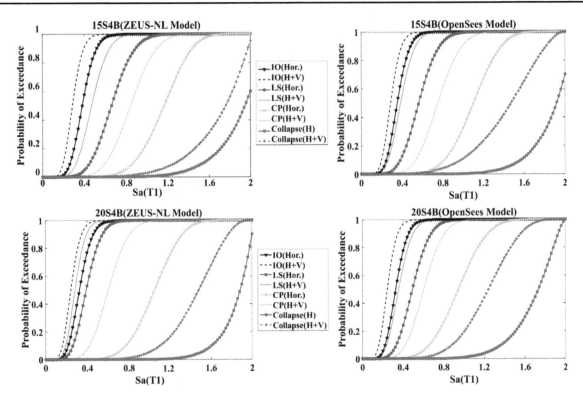

Fig. 19 Fragility curves for special RC-MRFs at the optimal HM

As discussed in Haselton et al. (2008) and FEMA-P695 (2009), the collapse capacity of structures and the calculated CMR could be strongly influenced by the frequency content (spectral shape) of the ground motions. Thus, in FEMA-P695, a spectral shape factor (SSF) is proposed to modify the CMR values. To this end, adjusted CMR (ACMR) value for each structure is assessed (Eq. 6):

$$ACMR = SSF \times CMR. \tag{6}$$

The SSF value for each structure is extracted based on the fundamental period (T_1), period-based ductility (μ_T) and the seismic design category from FEMA P-695 documentation. For seismic performance evaluation of the studied structures under concurrent horizontal–vertical excitations, on the basis of the 5 % probability of collapse under MCE and the composite uncertainty ($\beta_{TOT} = 0.5$) in collapse capacity, the acceptable ACMR for all building models takes the value of 2.28 according to FEMA-P695 recommendation. The final acceptance criteria for the reference structures are reported in Table 6. The results show the CMR, SSF and ACMR values for each archetypical design. Later on, the acceptable ACMRs are compared with the calculated values to see whether the structures could either pass or fail the criterion.

Focusing on the four reference structures, Table 6 shows that SMRFs have acceptable ACMR, while a disturbing trend becomes evident for the IMRF

buildings. The results show that in terms of coupled horizontal–vertical excitation, both 7S3B and 12S3B buildings have unacceptable ACMR, and surprisingly 7S3B would also fail for the horizontal excitation case while the 12-story frame passed the criteria marginally. As a result, the IMRFs do not attain the collapse performance required by FEMA P-695 methodology, and additional design requirement adjustments would be needed to improve the overall performance. It means that even the code-conforming design structures with acceptable level of ductility can be vulnerable to seismic excitations.

Comparison of the calculated ACMRs in terms of (H) and ($H + V$) excitations demonstrates that generally in all models, the safety margin against collapse reduces by including the vertical component of earthquake and the reductions are very remarkable and pronounced. Fortunately, the SMRF models could pass the collapse performance criteria for both types of excitation. It shows that their elements, being controlled by many detailing and capacity design requirements of the building code, limit possible failure modes.

Mean annual frequency (MAF) of collapse

Ground motion hazard curves for a typical highly seismic and populated region in the Middle-East is

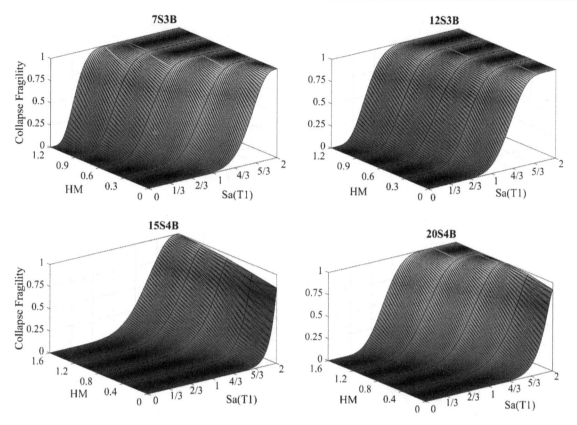

Fig. 20 Collapse fragility surface based on vector IM [$S_a(T_1)$, HM]

illustrated in Fig. 22 at the periods of 0.2, 1 and 3 s, and the required data at the fundamental period of structures are interpolated from these values. The buildings are assumed to be located at a site in Tehran, for which the hazard curve has been defined through probabilistic seismic hazard analysis (PSHA) by Gholipour et al. (2008). Another collapse metric, called the mean annual frequency (MAF) of collapse, is defined by integrating the seismic hazard curve and the fragility results. This metric describes how likely it is for collapses to occur, considering both the structural collapse capacity and the ground-shaking hazard in the studied region (Krawinkler et al. 2004; Liel 2008).

The mean annual frequency of collapse ($\lambda_{Collapse}$) is computed using Eq. (7) (Haselton and Deierlein 2007):

$$\lambda_{Collapse} = \int P[Sa_{Collapse} \leq x] \cdot |d\lambda_{IM}(x)|, \qquad (7)$$

where $P[Sa_{Collapse} \leq x]$ is the probability that x exceeds the collapse capacity (i.e., the probability that the building collapses when the ground motion intensity is x), and $\lambda_{IM}(x)$ is the mean annual frequency of the ground motion intensity exceeding x (i.e., a point on the seismic hazard curve). There are many ways to approximate Eq. (7). A closed-form solution is used to fit an exponential function to the hazard curve. To avoid error induced by fitting an exponential function, the PCHIP (piecewise cubic hermite interpolating polynomial) procedure is used to interpolate between the points (MathWorks 2015). In the next stage, the interpolated curves are implemented to complete the numerical integration required to evaluate Eq. (7) (Fig. 23).

Based on the extracted results in Fig. 23, the $\lambda_{Collapse(max)}$ in the intermediate MRFs are 2.1×10^{-5} and 1.86×10^{-4} collapse/year for the horizontal and combined $H + V$, respectively. However, these values are relatively smaller for the special MRFs under both types of excitations. The maximum values for the SMRFs are 2.3×10^{-6} and 4.5×10^{-5} collapse/year for the horizontal and combined $H + V$, respectively. It is worth mentioning that these results correspond to collapse return periods of 2475 years. Based on MAF calculation, SMRFs are in a higher confidence bounds of safety compared to the intermediate moment frames, while the effect of vertical ground motion is very significant for both groups and cannot be neglected.

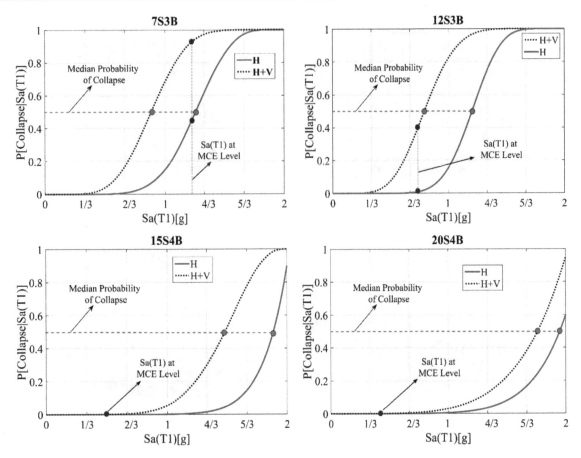

Fig. 21 Collapse fragility curves for intermediate and special RC-MRFs, illustrating key metrics for collapse performance

Table 6 Summary of final collapse margins and comparison to acceptance criteria

Reference structure	SSF	$S_a(T_1)_{@MCE}$	Loading type	$S_a(T_1)_{@50\%Col.}$	CMR	ACMR	Acceptance ACMR	Performance
7S3B	1.31	1.24	H	1.31	1.06	1.38	2.28	Fail
			$H + V$	0.88	0.71	0.93	2.28	Fail
12S3B	1.35	0.73	H	1.25	1.71	2.31	2.28	Marginal pass
			$H + V$	0.81	1.11	1.50	2.28	Fail
15S4B	1.61	0.50	H	1.85	3.70	5.96	2.28	Pass
			$H + V$	1.43	2.86	4.60	2.28	Pass
20S4B	1.61	0.37	H	1.91	5.16	8.31	2.28	Pass
			$H + V$	1.76	4.76	7.66	2.28	Pass

Conclusions and recommendations

In this study, the effect of vertical excitations on the seismic performance of intermediate and special RC-MRFs has been evaluated. It is important to emphasize that the main objective of this study was not to quantify numerical design values. The objective was rather to focus on the importance or otherwise of including vertical ground motion in design of RC buildings and its impact on the member and the structure levels. Hence, a large number of NL dynamic analyses were performed using fiber-based and concentrated plasticity approaches. The computational models were utilized in ZEUS-NL and OpenSees platforms to compare the results. The VGM was shown to be significant and should be included in the analysis when the proposed structure is located within 25 km of a seismic source.

The most important findings are summarized as follows:

- The vertical component of an earthquake tends to concentrate all its energy content in a narrow band, unlike the horizontal counterpart. This energy

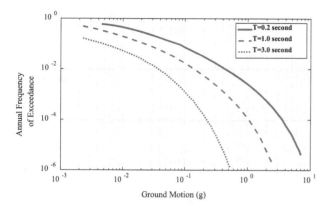

Fig. 22 Tehran (Iran) hazard curve at selected periods (Gholipour et al. 2008)

concentration can be very destructive for the (mid- and high)-rise RC-MRFs with vertical periods in the range of vertical components periods. Extracted ACMR values and fragility curves proved that the intermediate RC-MRFs are very vulnerable and need major revision in their design stage, while the SMRFs can well resist both horizontal and vertical seismic excitations.

- Based on the frequency content of the vertical ground motion, it can be concluded that structural failure modes from past earthquakes might be attributed to underestimating the effect of vertical acceleration in the design procedures, and there is an urgent need for the adoption of more realistic vertical spectra in future version of seismic design codes.

- Although the effect of axial force on shear capacity of the structural elements is an accepted fact and is proved in the current study, current seismic codes do not have a consensus on this effect, and different code equations might lead to different shear capacity estimations. Both the ACI-318 and ASCE-41 equations captured the shear strength degradation due to axial force. ASCE equation predictions could be considered as accurate, because

the strength reduction caused by ductility demand could be more significant than that by tension.

- As the V/H ratio increases, more fluctuations can be observed in the columns axial force. This phenomenon leads to a significant reduction in the shear capacity in the range of (15–30) %. This reduction in shear capacity of the vertical members increases the potential for shear failure.

- Geometric nonlinearities, in terms of the deformed configuration of the system, do not come into play in either IO or LS damaged states of the system. However, P-Δ effects due to higher interstory drifts of combined $H + V$ excitations do influence the response of the building in the region near collapse and must be taken into account.

- A new vector IM is proposed in the current study to predict accurate fragility results. One of the main advantages of the proposed IM is its hazard compatibility, in which a GM prediction model can be easily developed for the second parameter (HM), implementing the existing attenuation models with an arbitrary set of periods. In case a probabilistic seismic hazard analysis (PSHA) would be required, the calculations can be performed using Ln (HM) as IM in the same way as any single spectral acceleration value.

- The results presented in this study indicate that coupling horizontal and vertical ground excitations increases the ductility demand. Therefore, it is highly suggested that the conventional response spectrum (RS) be replaced with a multi-component RS in the next version of seismic design codes. Doing this will consider the effect of vertical ground excitations on the enhanced seismic demand and will provide a better understanding to structural designers.

Taking into account the above observations, the authors would like to recommend for the next version of seismic codes to make sure that the structures locate within 25 km

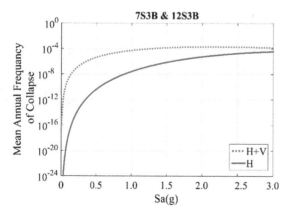

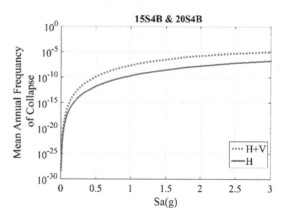

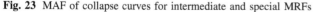

Fig. 23 MAF of collapse curves for intermediate and special MRFs

from the active faults be designed to the combined effect of horizontal and vertical ground motions.

References

ACI (2011) Building code requirements for structural concrete (ACI 318-11) and commentary, (ACI 318R-11). American concrete Institute, Farmington Hills

Ahmadi G, Mostaghel N (1978) On the stability of columns subjected to non-stationary random or deterministic support motion. J Earthq Eng Struct Dyn 3:321–326

Ahmadi G, Mostaghel N (1980) Stability and upper bound to the response of tall structures to earthquake support motion. J Earthq Eng Struct Mech 8:151–159

AIJ (I995) Preliminary reconnaissance report of the 1995 Hyogo-ken Nanbu Egthquake, Architectural Institute of Japan, April, p 216

Ancheta T, Bozorgnia Y, Darragh R, Silva WJ, Chiou B, Stewart JP, Atkinson GM (2012) PEER NGA-West2 database: a database of ground motions recorded in shallow crustal earthquakes in active tectonic regions. In: Proceedings, 15th World Conference on Earthquake Engineering

Ancheta TD, Darragh RB, Stewart JP, Seyhan E, Silva WJ, Chiou BSJ et al (2013) PEER NGA-West2 Database. PEER report 2013. Pacific Earthquake Engineering Research, Berkeley, USA, 3, 83

Anderson JC, Bertero VV (1977) Effect of gravity loads and vertical ground accelerations on the seismic response of multi-story frames. In: Proceedings of the World Conference on Earthquake Engineering, 5th Paper, Rome, Italy, pp 2914–2919

Antoniou S (1997) Shear assessment of R/C structures under combined earthquake loading. MSc Dissertation, ESEE, Imperial College, London, UK

ASCE (2010) Minimum design loads for buildings and other structures, ASCE standard ASCE/SEI 7-10. American Society of Civil Engineers, Reston

ASCE/SEI Seismic Rehabilitation Standards Committee (2013) Seismic rehabilitation of existing buildings (ASCE/SEI 41-13). American Society of Civil Engineers, Reston

Aschheim M, Moehle JP (1992) Shear strength and deformability of RC bridge columns subjected to inelastic displacements. Report UCB/EERC 92/04, University of California, Berkeley

Baker JW (2015) Efficient analytical fragility function fitting using dynamic structural analysis. Earthq Spectra 31(1):579–599

Bozorgnia Y, Campbell KW (2004) The vertical-to-horizontal response spectral ratio and tentative procedures for developing simplified V/H and vertical design spectra. J Earthq Eng 8(2):175–207

Bozorgnia Y, Campbell K, Niazi M (1999) Vertical ground motion: characteristics, relationship with horizontal component, and building-code implications. In: SMIP99 Seminar Proceedings, pp 23–49

Broderick BM, Elnashai AS, Ambraseys NN, Barr JM, Goodfellow RG, Higazy EM (1994) The Northridge (California) earthquake of 17 January 1994: observations, strong motion and correlative response analyses. ESEE research report No. 94/4, Imperial College, London

Button M, Cronin C, Mayes R (2002) Effect of vertical motions on seismic response of highway bridges. J Struct Eng 128(12):1551–1564

Calvi GM, Pinho R, Magenes G, Bommer JJ, Restrepo-Velez LF, Crowley H (2006) Development of seismic vulnerability assessment methodologies over the past 30 years. ISET J Earthq Technol 43(3):75–104

Chopra AK (1995) Dynamics of structures. Prentice Hall, New Jersey

Ditlevsen O, Madsen HO (1996) Structural reliability methods. Wiley, New York

Elgamal A, He L (2004) Vertical earthquake ground motion records: an overview. J Earthq Eng 8(05):663–697

Elnashai AS, Bommer JJ, Baron CI, Lee D, Salama AI (1995) Selected engineering seismology and structural engineering studies of the Hyogo-Ken Nanbu (Great Hanshin) earthquake of 17 January 1995. ESEE research report No. 95-2, Imperial College, London

Elnashai AS, Papanikolaou VK, Lee DH (2004) Zeus NL-A system for inelastic analysis of structures. CD-Release 04-01, Mid-America Earthquake Center, University of Illinois at Urbana-Champaign, Urbana, IL

Elwood KJ (2004) Modelling failures in existing reinforced concrete columns. Can J Civ Eng 31:846–859

Elwood KJ, Moehle JP (2005) Drift capacity of reinforced concrete columns with light transverse reinforcement. Earthq Spectra 21:71–89

FEMA P-695 (2009) Quantification of seismic performance factors. FEMA P-695 report, the Applied Technology Council for the Federal Emergency Management Agency, Washington, DC

FEMA 356 (2000) NEHRP guidelines for the seismic rehabilitation of buildings. Federal Emergency Management Agency, Washington, DC

FEMA P-440A (2009) Effects of strength and stiffness degradation on seismic response. Federal Emergency Management Agency, Washington, DC

Friedland I, Power M, Mayes R (1997) Proceedings of the FHWA/NCEER Workshop on the National Representation of Seismic Ground Motion for New and Existing Highway Facilities, National Center for Earthquake Engineering Research, Technical Report NCEER-97-0010, Burlingame

Gardoni P, Der Kiureghian A, Mosalam KM (2002) Probabilistic capacity models and fragility estimates for RC columns based on experimental observations. J Eng Mech 128(10):1024–1038

Gardoni P, Mosalam KM, Der Kiureghian A (2003) Probabilistic seismic demand models and fragility estimates for RC bridges. J. Earthq Eng 7(1):79–106

Gholipour Y, Bozorgnia Y, Rahnama M, Berberian M, Shojataheri J (2008) Probabilistic seismic hazard analysis, phase I—greater Tehran regions. Final report. Faculty of Engineering, University of Tehran, Tehran

Giovenale P, Cornell CA, Esteva L (2004) Comparing the adequacy of alternative ground motion intensity measures for the estimation of structural responses. Earthq Eng Struct Dyn 33:951–979

Haselton CB, Deierlein GG (2007) Assessing seismic collapse safety of modern reinforced concrete frame buildings. PEER Rep. 2007/08, PEER Center, University of California, Berkeley, CA

Haselton CB, Liel AB, Taylor Lange S, Deierlein GG (2008) Beam-column element model calibrated for predicting flexural response leading to global collapse of RC frame buildings. PEER Rep. 2007/03, PEER Center, University of California, Berkeley, CA

Ibarra LF, Medina RA, Krawinkler H (2005) Hysteretic models that incorporate strength and stiffness deterioration. Int J Earthq Struct Dyn 34(12):1489–1511

Iyengar RN, Sahia TK (1977) Effect of vertical ground motion on the response of cantilever structures. In: Proceedings of the Sixth

World Conference on Earthquake Engineering, New Delhi, India, pp 1166–1177

Iyengar RN, Shinozuka M (1972) Effect of self-weight and vertical accelerations on the seismic behaviour of tall structures during earthquakes. J Earthq Eng Struct Dyn 1:69–78

Munshi JA, Ghosh SK (1998) Analyses of seismic performance of a code designed reinforced concrete building. J Eng Struct 20(7):608–616

Jeon JS (2013) Aftershock vulnerability assessment of damaged reinforced concrete buildings in California. Ph.D. Thesis, Georgia Institute of Technology, USA

Kim SJ, Holub CJ, Elnashai AS (2011) Analytical assessment of the effect of vertical earthquake motion on RC bridge piers. J Struct Eng 137(2):252–260

Kramer SL, Arduino P, Shin HS (2008) Using OpenSees for performance-based evaluation of bridges on liquefiable soils, PEER Report 2008/07. Pacific Earthquake Engineering Research Center, University of California, Berkeley

Krawinkler H, Miranda E, Bozorgnia Y, Bertero VV (2004) Chapter 9: Performance based earthquake engineering. Earthquake engineering: from engineering seismology to performance-based engineering. CRC Press, Florida

LeBorgne MR. (2012) Modeling the post shear failure behaviour of reinforced concrete columns. Ph.D. dissertation, University of Texas at Austin, Austin, Texas

Lee DH, Elnashai AS (2001) Seismic analysis of RC bridge columns with flexure-shear interaction. J Struct Eng 127(5):546–553

Lee DH, Elnashai AS (2002) Inelastic seismic analysis of RC bridge piers including flexure–shear–axial interaction. Struct Eng Mech 13(3):241–260

Liel AB (2008) Assessing the collapse risk of California's existing reinforced concrete frame structures: metrics for seismic safety decisions. Ph.D. dissertation, Stanford University

Lowes LN, Altoontash A (2003) Modeling reinforced-concrete beam-column joints subjected to cyclic loading. ASCE J Struct Eng 129(12):1686–1697

Luco N, Cornell CA (2007) Structure-specific scalar intensity measures for near-source and ordinary earthquake ground motions. Earthq Spectra 23(2):357–392

Mackie K, Stojadinovic B (2001) Probabilistic seismic demand model for California highway bridges. J Bridge Eng 6:468–481

Mander JB, Priestley MJN, Park R (1988) Theoretical stress-strain model for confined concrete. J Struct Eng 114(8):1804–1826

MathWorks Inc (2015) MATLAB: the language of technical computing. Desktop tools and development environment, Version 2015a

McKenna F (2014) Open system for earthquake engineering simulation, OpenSees, version 2.4. 4 MP [Software], Pacific Earthquake Engineering Research Center

Mostaghel N (1974) Stability of columns subjected to earthquake support motion. J Earthq Eng Struct Dyn 3(4):347–352

Newmark NM, Blume JA, Kapur KK (1973) Seismic design spectra for nuclear power plants. J Power Div 99(2):287–303

Pacific Earthquake Engineering Research Center (2015) PEER NGA-WEST2 database flat file

Padgett JE, Nielson BG, DesRoches R (2008) Selection of optimal intensity measures in probabilistic seismic demand models of highway bridge portfolios. Earthq Eng Struct Dyn 37:711–725

Papazoglou AJ, Elnashai AS (1996) Analytical and field evidence of the damaging effect of vertical earthquake ground motion. Earthq Eng Struct Dyn 25(10):1109–1137

PEER/ATC-72-1 (2010) Modeling and acceptance criteria for seismic design and analysis of tall buildings. Applied Technology Council in cooperation with the Pacific Earthquake Engineering Research Center, Redwood City

Priestley MJN, Verma R, Xiao Y (1994) Seismic shear strength of reinforced concrete columns. J Struct Eng 120(8):2310–2329

Sezen H (2002) Seismic response and modeling of reinforced concrete building columns. Ph.D. thesis, Department of Civil and Environmental Engineering, University of California, Berkeley, California

Shin M, LaFave JM (2004) Modeling of cyclic joint shear deformation contributions in RC beam-column connections to overall frame behavior. Struct Eng Mech 18(5):645–669

Vamvatsikos D, Cornell CA (2002) Incremental dynamic analysis. Earthq Eng Struct Dyn 31(3):491–514

Vecchio FJ, Collins MP (1986) The modified compression field theory for reinforced concrete elements subjected to shear. ACI Struct J 83(2):219–231

Watanabe F, Ichinose T (1992) Strength and ductility of RC members subjected to combined bending and shear. In: Hsu TTC, Mau ST (eds) Concrete shear in earthquake. Elsevier Applied Science, New York, pp 429–438

Wen YK, Ellingwood BR, Bracci JM (2004) Vulnerability functions, Technical Rep. DS-4, Mid-America Earthquake Center, University of Illinois at Urbana-Champaign

Zareian F, Medina RA (2010) A practical method for proper modeling of structural damping in inelastic plane structural systems. J Comput Struct 88:45–53

13

NDT evaluation of long-term bond durability of CFRP-structural systems applied to RC highway bridges

Kenneth C. Crawford[1]

Abstract The long-term durability of CFRP structural systems applied to reinforced-concrete (RC) highway bridges is a function of the system bond behavior over time. The sustained structural load performance of strengthened bridges depends on the carbon fiber-reinforced polymer (CFRP) laminates remaining 100 % bonded to concrete bridge members. Periodic testing of the CFRP–concrete bond condition is necessary to sustain load performance. The objective of this paper is to present a non-destructive testing (NDT) method designed to evaluate the bond condition and long-term durability of CFRP laminate (plate) systems applied to RC highway bridges. Using the impact-echo principle, a mobile mechanical device using light impact hammers moving along the length of a bonded CFRP plate produces unique acoustic frequencies which are a function of existing CFRP plate–concrete bond conditions. The purpose of this method is to test and locate CFRP plates de-bonded from bridge structural members to identify associated deterioration in bridge load performance. Laboratory tests of this NDT device on a CFRP plate bonded to concrete with staged voids (de-laminations) produced different frequencies for bonded and de-bonded areas of the plate. The spectra (bands) of frequencies obtained in these tests show a correlation to the CFRP–concrete bond condition and identify bonded and de-bonded areas of the plate. The results of these tests indicate that this NDT impact machine, with design improvements, can potentially provide bridge engineers a means to rapidly evaluate long lengths of CFRP laminates applied to multiple highway bridges within a national transportation infrastructure.

Keywords CFRP plates · Non-destructive testing · Bond · Impact-echo · Frequency · Bridges

Introduction

The load performance of CFRP-strengthened RC highway bridges is a function of the bond-interface behavior in the CFRP laminate–epoxy–concrete structural system. It is necessary to periodically evaluate the bond condition of the CFRP laminates applied to bridges to monitor potential changes (deteriorations) in bridge load performance. For transportation infrastructure owners, and in particular owners of highway bridges strengthened with CFRP systems, one primary concern is the long-term durability of the CFRP systems and sustainment of the designed bridge load capacity. Bridge owners are faced with two primary questions: (1) how do CFRP structural systems on RC highway bridges perform over extended time (15+ years) under the influence of frequent cyclic loading (heavy truck traffic), moisture, and freeze-thaw cycles? (2) How can accurate field data be obtained on CFRP-plate bond condition to evaluate changes in load performance and to establish cost-effective maintenance procedures? This concern is magnified when a national highway system has a large number of strengthened bridges and the load performance of the bridges is critically dependent on the CFRP plate–concrete bond performance, i.e., absence of de-bonding. CFRP plates with de-laminations can potentially produce deterioration in bridge performance (Shih et al. 2003). The NDT concept presented in this paper is one method for rapidly evaluating the bond condition of CFRP

✉ Kenneth C. Crawford
 ken.crawford69@wabash.edu

[1] Crawford Technologies and Applications LLC,
 Crawfordsville, IN 47933, USA

plates applied to multiple bridges in a national highway system. A comprehensive CFRP-strengthened bridge inspection and maintenance program with periodic testing and evaluation of the CFRP plate bond condition is necessary to support and sustain long-term bridge performance.

Background

The Republic of Macedonia carried out a NATO-funded program in 2001 to strengthen 19 highway bridges on the M2 and M1 highways on European Corridor 8 connecting Albania to Bulgaria through northeastern Macedonia. CFRP-plate structural systems were applied to multiple bridges on the M2, and two bridges on the M1, to increase bridge load capacity for heavy military vehicle traffic. This FRP application significantly increased the flexural bending moment of bridge structural members. A total of 14,600 m of CFRP plates were applied to the 19 bridges having a bonded area of 1318 square meters (Crawford and Nikolovski 2007). Consequently, the condition of CFRP laminate–concrete bond becomes a critical element in sustaining long-term bridge load performance. After 15 years the condition of the CFRP–concrete bond on the 19 bridges is an open question: are the CFRP plates still 100 % bonded to the bridge structural members? Is the current bridge load performance sustaining the original designed CFRP-strengthened flexural bending moment? The challenge is having an effective field testing process to evaluate the CFRP-plate bond condition to determine if any changes in bridge performance have occurred. Since 2006 an additional 82 bridges in Macedonia have been strengthened and refurbished with FRP systems in varying applications. 46 bridges in the Ukraine, nine bridges in Montenegro, and six bridges in Kosovo have also been refurbished and strengthened. The most recent contract in Macedonia was awarded in December 2013, to repair and strengthen 16 bridges on the M1 (European Corridor 10) between Skopje and 50 km south to Veles. To sustain the designed load performance on this large number of CFRP-strengthened bridges, a reliable field testing procedure in bridge inspection and maintenance is needed to periodically verify bond condition and associated bridge load performance.

Scope of strengthened-bridge applications requiring CFRP-bond testing

The 19 bridges described above have varying degrees of CFRP laminations applied to the structural members of different types of slab and girder bridges. Figures 1 and 2

Fig. 1 Bridge B2-N on M1—938 m

Fig. 2 Bridge B36 on M2—1032 m

illustrate the lengths of CFRP plates applied to two slab bridges, each having about 1 km of CFRP material. Table 1 describes, as an example, the length and area of CFRP plate material applied on ten of the 19 bridges.

One example of the degree of CFRP strengthening applied to the 19 slab and girder bridges on the M1 and M2 is illustrated by the bridge B2-N, Fig. 1, at km marker 10+7 near Kumanovo on the M1. B2-N is a 46 m bridge in three spans, 10.2 m wide, 66 cm thick, constructed in 1962 in accordance with YU Regulations PTP-5 for an M-25 vehicle. For this bridge the ultimate bending moment before strengthening was 399.8 kNm. The required ultimate bending moment after strengthening was 659.2 kNm, a 65 % increase in the bending moment. 938 m of CFRP plates were applied to the bridge to achieve the 659 kNm bending moment (Crawford 2008). To sustain this bending moment 100 % of the plates must be bonded to the bridge. It therefore becomes necessary to periodically evaluate the condition of the laminate bond to verify the designed load performance is still being sustained by the applied CFRP system.

Table 1 Length/area of applied CFRP plates on ten M2 and M1 bridges

Selected CFRP-strengthened bridges on the M2 and M1					Total CFRP plate length (m) and area (m^2)		
Bridge no.	Location on M2/M1	Bridge type	Bridge length (m)	No. spans	No. plates	Plate length	Plate area
B7	14+027	Girder	120	6	72	1478	116.1
B11	21+876	Slab	10	1	26	198	29.6
B18	38+444	Slab	36	1	54	218	32.6
B22	41+786	Slab	30	1	60	1308	196.3
B28	49+631	Girder	50	3	60	1210	100.6
B35	66+058	Girder	52	3	20	346	41.2
B36	67+409	Slab	21	2	80	1032	82.8
B37	68+452	Girder	17	1	36	415	34.4
B39	71+211	Girder	85	4	48	1778	146.2
B2-N	10+07 M1	Slab	46	4	96	938	112.5

Significance of research

Owners using CFRP-structural systems on large numbers of bridges within a national transportation system have an ongoing requirement to evaluate the bond condition of FRP composites applied to RC highway bridges in order to monitor long-term bridge load performance. While the impact-echo method has been effective in testing concrete structures for many years, its application for evaluating CFRP-structural systems applied to bridges is not fully developed and used in the field. The development of the mobile impact-echo machine presented in this paper expands the application of NDT procedures to evaluate bridge CFRP plate–concrete bond conditions in the field.

Impact-echo NDT evaluation of CFRP–concrete bond

Among available methods of non-destructive testing (NDT) to evaluate the condition of FRP material bonded to concrete structures, among which are thermal imaging, high-frequency radiation, and laser vibrometry, the use of the impact-echo method presented provides an effective and low-cost means of rapidly assessing the bond condition of the CFRP laminate–epoxy–concrete tri-layered structure applied to large numbers of strengthened bridges. The advantage of this method over the three methods above is ease and speed of testing long lengths of CFRP plates.

Impact-echo principle

The concept of impact-echo testing, invented by the US National Bureau of Standards (NBS) in the mid-1980s, applies acoustic methods for non-destructive evaluation of concrete and masonry structures. The principle of the impact-echo method involves transmission of energy of a mechanical impact from a small steel sphere into the mass of a structure to generate transient sinusoidal stress (sound) waves. Low-frequency impact stress waves propagating into a structure are reflected by internal surfaces and flaws (Sansalone and Streett 1998). The patterns present in the reflected waveforms and spectra provide information about the existence and locations of structural flaws, e.g., de-bonding. Multiple reflections of stress waves between the impact surface and flaws give rise to transient resonances, which can be identified in the frequency spectrum, and used to evaluate the integrity of the structure. The resulting wave displacement signals in the time domain are transformed into the frequency domain in which plots of amplitude versus frequency (spectra) are obtained. Surface displacements caused by these wave reflections are recorded by a transducer located adjacent to the point of impact. Impact-echo tests on concrete structures, such as CFRP plates applied to bridge structural members, produce distinctive waveforms and frequencies, in which dominant patterns in the number and distribution of peaks in the spectra are easily identified. Voids and de-laminations (flaws) in the CFRP–concrete structure change impact waveform patterns (frequency spectra) and provide qualitative and quantitative information about the existence and location of the structural deficiencies (Ray et al. 2007).

The quality of impact-echo testing and the data obtained from the CFRP–concrete structure is a function of the impact hammer configuration and the sinusoidal waveform it provides. The objective is to achieve maximum waveform resolution with optimized hammer sphere size and impact force. In a mechanical system the impact of a small steel sphere can be represented as a unit impulse function with impact force $g(t)$. The impact energy $I(\tau)$, Eq. 1, is the integral of the impact force $g(t)$ over the impulse time interval $(t_1 - t_0)$. If force $g(t) = c$ and $t_0 = 0$ with $t_1 < 180$ µs such that $0 < \tau < 180$ µs, the integral of $I(\tau)$ from t_0 to t_1 becomes the total impact energy.

$$I(\tau) = \int_{t_0}^{t_1} g(t)\mathrm{d}t = 2\tau c = 1 \tag{1}$$

The unit impulse is the total hammer impact energy transmitted into the CFRP–concrete bonded structure. The input energy $I(\tau) = c = 1/2(t_1 - t_0)$. For the NDT device used in these tests, the force, where $g(t) = c$, of each hammer impact on the surface of the CFRP plate, was equal to 5 N (about 1 lb). The short duration of a 5-N impact force is sufficient for detecting flaws (de-bonding) in CFRP plate–concrete bond structure.

Fourier transform of impact signal

The impulse signal generated by the hammer impact on the surface of the CFRP plate produces a complex sinusoidal waveform in the time domain with a unique frequency that has amplitude and phase. The objective of the test is to identify the waveform frequency of the impact and to determine how the frequency of the impacts changes over the length of the CFRP plate. Given two sinusoidal waveforms t_1 and t_2, each t represents a different time and position on the plate in the hammer impact sequence. If t_1 is the signal of a bonded plate and t_2 is that of a de-bonded plate, each sinusoidal waveform signal t will have a different time phase and unique frequency. The equation for the sinusoidal signal t in the time domain is $f(t) = cos\ \omega t + i\ sin\ \omega t$, where $\omega = 2\pi f$ is the angular frequency in radians/sec. Using the Fourier transform in Eq. 2 the frequency $F(\omega)$ of the sinusoidal signal $f(t)$ is found by integrating $f(t)$ in its exponential form $e^{-i\omega t}$ (Euler's formula) from T_1 to T_2.

$$F(\omega) = \int_{T_1}^{T_2} f(t)e^{-i\omega t}\mathrm{d}(t) \tag{2}$$

If $f(t)$ from T_1 to T_2 is the duration of impulse generated in the time domain then the time signals t_1 and t_2 in the Fourier transform become the frequency signals f_1 to f_2. If f_1 and f_2 are two different frequencies such that if $f_1 < f_2$, then the impact signals t_1 and t_2 indicate two different bond conditions, e.g., de-bonded versus bonded CFRP plates. With the Fourier transform providing the frequencies of the time signals, changes in frequencies of the impact signals as the NDT device moves over the CFRP plate can be identified. If the impact frequencies across the CFRP plate remain constant, the plate is fully bonded. Any changes to lower frequencies can indicate some plate de-bonding has occurred and may be present over the length of the CFRP plate.

Mobile impact-echo machine

The NDT impact machine, Fig. 3, is a four-wheel mobile device consisting of two spring-activated levers mounted 5 cm apart in an enclosed frame. One actuator wheel alternately lifts each lever 10 mm to drop an impact bolt on the CFRP plate surface. Each lever is driven by an adjustable-tensioned band (spring) to obtain a consistent 5 N impact force along the length of the plate. The impacts occur alternately along two parallel tracks on the plate (Crawford 2011). The impact impulse energy is produced by a small steel sphere (head of a carriage bolt) mounted on the oscillating lever. The size of the sphere is optimized to produce maximum input energy into the CFRP plate surface. The resin matrix of the CFRP plate is a softer material than the concrete and does not produce as sharp a stress wave (sound) as an impact on a concrete surface. This means the system for detection and recording the impact signal must be more sensitive (responsive) to detect changes in frequency on the CFRP plate as bond conditions with the concrete change. Frequency differences for bonded versus de-bonded areas are generally smaller for changes in CFRP-plate bonding than for flaws in solid concrete structures.

CFRP test plate

To test the impact machine, a 10 cm by 3.0 m Sika CarboDur S1012 rigid plate, 1.2 mm thick, Fig. 4, was fixed with double-backed adhesive tape (in place of epoxy) on a level concrete surface. The bonding adhesive was spaced to create 10–40 cm delaminated (de-bonded) areas between the plate and the concrete. In Fig. 4 the de-bonded areas are marked in gray on the plate surface, showing their relationship to the bonded areas along the length of the test plate. The frequencies generated by the multiple impacts on the CFRP test plate, Fig. 4, are shown on two horizontal lines, each with sets of vertical lines. Each vertical line is a single impact point with a unique frequency indicating a bonded plate on the 3.26 kHz line and a de-bonded area of the plate on the 2.88 kHz line.

Testing protocol

The objective of this testing protocol is to verify the concept of applying the NDT impact-echo technique to identify de-bonding of the epoxy-bond structure between CFRP laminates and concrete bridge structural members. The goal is to determine the feasibility of using this testing procedure in the field on CFRP-strengthened bridges. The testing protocol was conducted in two phases: (1)

Fig. 3 Mobile impact machine
with dual light impact hammers

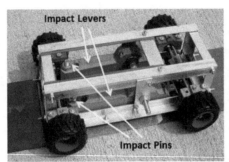
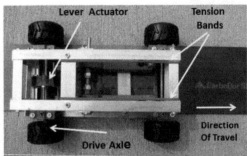

Fig. 4 CFRP test plate with
point-of-impact frequencies

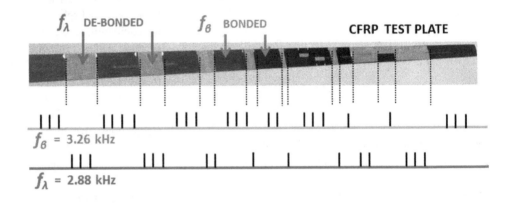

performing the impact-echo evaluation on the CFRP test plate and recording the resulting acoustic frequencies from the impacts. (2) Analyzing the recorded time-domain signals on the oscilloscope for pulse waveform structure and on the spectrum analyzer in the frequency domain for frequency variations and frequency components of the impact signals.

Test procedure

The mobile impact-echo machine shown in Fig. 3 was moved (rolled) along the length of the CFRP test plate, Fig. 4, at a constant speed. The two impact hammer levers were lifted alternately by an actuator wheel. With each drop of the lever the impact pin produced an impact stress wave (acoustic sound) every 3 cm, with the machine producing 33 impacts per meter across the surface of the CFRP plate. Figure 5 is a schematic of the mobile device producing multiple impact stress waves with successive impacts on the CFRP plate bonded to the concrete bridge structural member. Each impact produces a single acoustic signal with a unique frequency. The frequencies produced are a function of the bond condition at that specific point on the CFRP plate.

Impact 1 on a bonded plate in Fig. 5 produces a stress wave with frequency f_β. Impact 2 on a de-bonded area of the plate produces a stress wave with a lower frequency f_λ. Multiple runs with the impact device were conducted over

the length of the CFRP test plate. The impact signals were received through a transducer (microphone) mounted on the mobile device and recorded on a broad-band receiver for frequency spectrum analysis.

Test results

With the impact machine producing a series of 99 impacts along the length of the 3.0 m CFRP test plate the change in frequencies between the two CFRP-bond states (bonded vs. de-bonded) was distinct and measurable. This means the varying bond conditions of the CFRP plate did produce different impact frequencies. Test results with successive impacts on the CFRP test plate indicated voids (plate delaminations) were present when $f_\lambda < f_\beta$. The recorded impact signals were fed into a Tektronix 2712 (1 kHz to 1.86 GHz) spectrum analyser to display the range of frequencies produced by impacts over the bonded and de-bonded areas of the CFRP test plate. To do this the recorded signals were connected into the analyser preamplifier using a 2-kHz radio frequency (RF) center frequency F_c providing a reference marker for the input signals. The analyser ten-division horizontal scan was set to 1 kHz/division to provide a 10 kHz scanning span for the input frequency spectrum. A bandwidth resolution of 300 kHz enabled the recorded impact signal waveforms to be displayed on the spectrum analyser as smaller envelopes right of the F_c center frequency. The frequency spectra of the

Fig. 5 Schematic of stress waves produced by impact machine

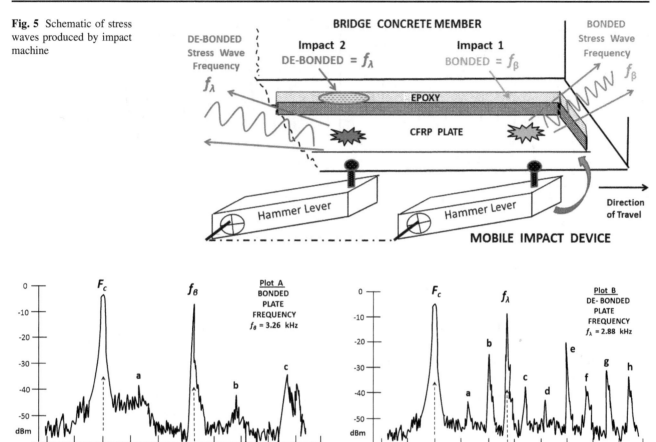

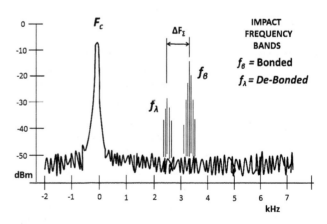

Fig. 6 Frequency spectrum display of bonded and de-bonded frequencies

impact signals are shown in plots A and B in Fig. 6. Both the bonded frequencies with primary peaks of 3.26 kHz and the de-bonded frequencies with primary peaks of 2.88 kHz appeared as clear and distinct frequencies with a measurable separation from the center frequency F_c in the range from 2.5 to 3.5 kHz. In these tests the amplitudes of the bonded versus de-bonded signals differed by -5 to -10 dBm. Smaller signal peaks right of center frequency F_c are positive harmonics of the impact frequencies. The harmonics shown on the analyser as peaks **a** through **g**, Fig. 6, are frequency components of the impact signals.

For this impact sequence the spectrum analyser indicated the frequency f_β for bonded CFRP plate was 3.26 kHz, and the frequencies for de-bonded areas of the CFRP plate fell in a narrow frequency range around 2.88 kHz, Fig. 6. The impact frequencies f_λ for the de-bonded areas were 10–12 % lower than the impact frequencies f_β for bonded areas. It was noted the range of frequencies f_λ over all de-bonded areas was consistent and remained in range of 10 % below the bonded frequencies. As the voids under the test plate became smaller (<10 cm) the frequency difference between f_λ and f_β became more

difficult to detect, possibly a function of the sensitivity of the recording system used in the tests. It was observed the frequencies for both the bonded and de-bonded areas varied slightly for each impact but remained generally within a 15 % range of the primary (strongest) signal frequency.

In Fig. 7 two separate frequency spectra bands are plotted for each set of recorded bonded and de-bonded

Fig. 7 Frequency bands for bonded and de-bonded impact signals

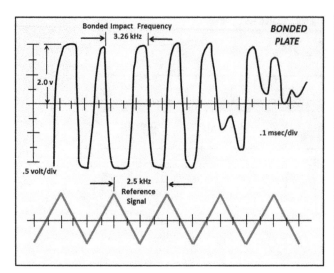

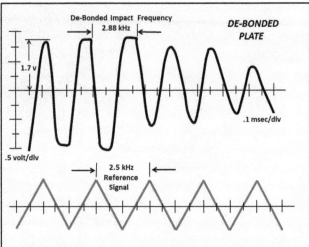

Fig. 8 Oscilloscope plots of bonded and de-bonded impact signal waveforms

impact signal frequencies. The majority of the bonded impact frequencies fell into a narrow frequency range between 3.10 and 3.40 kHz, indicated with frequency band f_β. The de-bonded frequencies also fell within a narrow frequency range between 2.6 and 2.9 kHz, shown by frequency band f_λ. The separation between the bonded and de-bonded frequency bands averaged approximately 300 Hz, indicated by ΔF_Σ in Fig. 7.

The recorded signals were also displayed on a Tektronix 475 dual-trace oscilloscope to analyze the waveform and structure of the impact signals. The input signals were displayed on the top trace of the oscilloscope, plotted in Fig. 8, with a 2.5-kHz sawtooth signal on the bottom trace providing a frequency reference scale. Both bonded and de-bonded impact signals had clear sinusoidal waveforms resembling more irregular square waves than pure sine waves. Both bonded and de-bonded waveforms remained consistent for successive impacts across the length of the CFRP test plate.

Discussion

The difference between bonded frequency f_β and de-bonded frequency f_λ is represented by ΔF_Σ, Fig. 7. The frequency difference ΔF_Σ is a function of the condition of the CFRP–concrete bond and expresses the degree of de-bonding of the CFRP plate over the length of the plate. When $\Delta F_\Sigma = 0$, then de-bonded frequencies f_λ are not present, meaning the CFRP plate is fully bonded. If $\Delta F_\Sigma > 0.3$ kHz then some de-bonding of the CFRP plate may have occurred. The degree of de-bonding can be quantified in terms of number of de-bonded frequencies occurring per meter over the length of the CFRP plate. This is accomplished by counting the

number of impacts per meter producing f_λ frequencies using a distance timing signal on the NDT device. The number of timing signals from the starting point will determine the location of the de-bonded areas on the plate. As the number of f_λ frequencies are recorded with their locations an accurate diagram of the de-bonded areas of the CFRP plate can be plotted and quantified. The data produced with this impact-echo method provide the bridge engineer an extensive picture of the CFRP plate bond condition. As bond condition evaluation tests are performed in periodic bridge maintenance protocols, the long-term performance and bond durability of the CFRP-structural systems applied to RC highway bridges can be determined, providing data to more accurately plan and program bridge maintenance costs.

Conclusion

The following points can be concluded from the tests presented in this paper:

1. Use of the NDT impact-echo method presented in this paper may be effective in evaluating CFRP laminate bond condition on RC bridge structural members.
2. Laboratory tests of the impact-echo machine with associated signal analysis show a correlation between signal frequency and CFRP-laminate bond condition.
3. The impact-echo method presented may be an effective means for rapidly evaluating the bond condition of long lengths of CFRP plates applied to large numbers of highway bridges.
4. Much research is needed on developing methods for signal processing and analysis of CFRP-laminate impact data points.

5. The impact tests presented in this paper indicate the impact-echo method can potentially identify changes in bridge load performance resulting from de-bonded CFRP laminate systems.

To monitor load performance of RC highway bridges strengthened with CFRP-structural systems, there is a growing need for effective field procedures to evaluate the bond structural condition of CFRP plates applied to extensive numbers of highway bridges. The NDT device presented in this paper using the technology of impact-echo testing, with further development, can have direct application to current bridge testing and maintenance programs for owners of CFRP-strengthened bridges. This NDT method can potentially provide bridge engineers a practical procedure to evaluate large areas of CFRP plates applied to multiple highway bridges in a national highway.

Acknowledgments The author thanks Wabash College for their support in the research for this paper.

Compliance with ethical standards

Conflict of interest The author declares he has no competing or conflicts of interest.

References

Crawford K (2008) Measuring performance of concrete bridges strengthened with FRP-structural systems. Advanced composite materials in bridges and structures (ACMBS-V), paper 53

Crawford K (2011) Non-destructive testing of FRP-structural systems applied to concrete bridges, international symposium on nondestructive testing of materials and structures, paper 102

Crawford K, Nikolovski T (2007) The application of ACI 440 for FRP system design to strengthen 19 concrete highway bridges on European corridor 8". In: 8th international conference on fiber reinforced polymer for reinforcement of concrete structures, paper 15–4, pp 570–571

Ray BC, Clegg DW, Hasan ST (2007) Evaluation of defects in FRP composites by NDT techniques. J Reinf Plast Compos 26:1187–1192

Sansalone M, Streett W (1998) The impact-echo method. NDTnet 3(2):3–7

Shih J, Tann D, Hu C, Delpak R, Andreou E (2003) Remote sensing of air blisters in concrete-FRP bond layer using IR thermography. Int J Mater Prod Technol 19:174–187

Permissions

All chapters in this book were first published in IJASE, by Springer; hereby published with permission under the Creative Commons Attribution License or equivalent. Every chapter published in this book has been scrutinized by our experts. Their significance has been extensively debated. The topics covered herein carry significant findings which will fuel the growth of the discipline. They may even be implemented as practical applications or may be referred to as a beginning point for another development.

The contributors of this book come from diverse backgrounds, making this book a truly international effort. This book will bring forth new frontiers with its revolutionizing research information and detailed analysis of the nascent developments around the world.

We would like to thank all the contributing authors for lending their expertise to make the book truly unique. They have played a crucial role in the development of this book. Without their invaluable contributions this book wouldn't have been possible. They have made vital efforts to compile up to date information on the varied aspects of this subject to make this book a valuable addition to the collection of many professionals and students.

This book was conceptualized with the vision of imparting up-to-date information and advanced data in this field. To ensure the same, a matchless editorial board was set up. Every individual on the board went through rigorous rounds of assessment to prove their worth. After which they invested a large part of their time researching and compiling the most relevant data for our readers.

The editorial board has been involved in producing this book since its inception. They have spent rigorous hours researching and exploring the diverse topics which have resulted in the successful publishing of this book. They have passed on their knowledge of decades through this book. To expedite this challenging task, the publisher supported the team at every step. A small team of assistant editors was also appointed to further simplify the editing procedure and attain best results for the readers.

Apart from the editorial board, the designing team has also invested a significant amount of their time in understanding the subject and creating the most relevant covers. They scrutinized every image to scout for the most suitable representation of the subject and create an appropriate cover for the book.

The publishing team has been an ardent support to the editorial, designing and production team. Their endless efforts to recruit the best for this project, has resulted in the accomplishment of this book. They are a veteran in the field of academics and their pool of knowledge is as vast as their experience in printing. Their expertise and guidance has proved useful at every step. Their uncompromising quality standards have made this book an exceptional effort. Their encouragement from time to time has been an inspiration for everyone.

The publisher and the editorial board hope that this book will prove to be a valuable piece of knowledge for researchers, students, practitioners and scholars across the globe.

List of Contributors

Araliya Mosleh
Department of Civil Engineering, Faculty of Engineering, University of Aveiro, 3810-193 Aveiro, Portugal

Mehran S. Razzaghi
Department of Civil Engineering, Faculty of Engineering, Islamic Azad University, Qazvin Branch, Qazvin, Iran

José Jara
Department of Civil Engineering, Faculty of Engineering, University Michoacana de San Nicolas de Hidalgo, Morelia, Mexico

Humberto Varum
CONSTRUCT-LESE, Department of Civil Engineering, Faculty of Engineering, University of Porto, 4200-465 Porto, Portugal

M. Mastali
Department of Civil Engineering, ISISE, Minho University, Campus de Azurem, 4800-058 Guimaraes, Portugal

R. Vahdani
Faculty of Civil Engineering, Semnan University, Seman, Iran

A. Kheyroddin
Faculty of Civil Engineering, Semnan University, Seman, Iran

Department of Civil Engineering and Applied Mechanics, University of Texas, Arlington, TX, USA

B. Samali
School of Civil and Environmental Engineering, University of Western Sydney, Sydney, Australia

M. S. Razzaghi and M. Khalkhaliha
Qazvin Branch, Islamic Azad University, Qazvin, Iran

A. Aziminejad
Science and Research Branch, Islamic Azad University, Tehran, Iran

Vahid Tahouneh
Young Researchers and Elite Club, Islamshahr Branch, Islamic Azad University, Islamshahr, Iran

Mohammad Hasan Naei
School of Mechanical Engineering, Tehran University, Tehran, Iran

Adel ElSafty
Civil Engineering Department, University of North Florida, Jacksonville, FL, USA

Matthew K. Graeff
Segars Engineering, 1200 Five Springs Rd., Charlottesville, VA, USA

Georges El-Gharib
Johnson, Mirmiran & Thompson, Inc., Lake Mary, FL, USA

Ahmed Abdel-Mohti
Civil Engineering Department, Ohio Northern University, Ada, OH, USA

N. Mike Jackson
Department of Civil Engineering and Construction Management, Georgia Southern University, Statesboro, GA, USA

Alireza Gholamhoseini
The University of Canterbury, Christchurch, New Zealand

Rony Kar and Sujit Kumar Dalui
Department of Civil Engineering, Indian Institute of Engineering Science and Technology, Shibpur, Howrah, India

Bhavana Valeti, Samit Ray-Chaudhuri and Prishati Raychowdhury
Indian Institute of Technology Kanpur, Kanpur, India

Alka Y. Pisal
Department of Civil Engineering, D. Y. Patil College of Engineering, Akurdi, Pune 411044, India

R. S. Jangid
Department of Civil Engineering, Indian Institute of Technology Bombay, Powai, Mumbai 400076, India

Mohammad Reza Sheidaii
Department of Civil Engineering, College of Engineering, Urmia University, Urmia, Iran

Mehrzad TahamouliRoudsari and Mehrdad Gordini
Department of Civil Engineering, Kermanshah Branch, Islamic Azad University, Kermanshah, Iran

Boshra El-Taly
Lecturer in Civil Engineering Department, Faculty of Engineering, Minoufia University, Shibin Al Kawm, Egypt

Ehsan Noroozinejad Farsangi
International Institute of Earthquake Engineering and Seismology (IIEES), Tehran, Iran

Abbas Ali Tasnimi
Tarbiat Modares University, Tehran, Iran

Kenneth C. Crawford
Crawford Technologies and Applications LLC, Crawfordsville, IN 47933, USA

Index

Printed in the USA
CPSIA information can be obtained
at www.ICGtesting.com
JSHW051444221024
72173JS00006B/1570